Eye Tracking Methodology

Andrew Duchowski

Eye Tracking Methodology

Theory and Practice

Second Edition

 Springer

Andrew Duchowski, BSc, PhD
Department of Computer Science
Clemson University
Clemson, SC 29634
andrewd@cs.clemson.edu

ISBN 978-1-84628-608-7   e-ISBN 978-1-84628-609-4

British Library Cataloguing in Publication Data
A catalogue record for this book is available from the British Library

Library of Congress Control Number: 2006939204

Printed on acid-free paper.

9 8 7 6 5 4 3 2 1

Springer Science+Business Media
springer.com

Preface to the Second Edition

Since the writing of the first edition, several important advancements in the field of eye tracking have occurred in the span of just a few short years. Most important, eye tracking technology has improved dramatically. Due to the increased speed of computer processors and improved computer vision techniques, eye tracking manufacturers have developed devices falling within the fourth generation of the following technological taxonomy.

1. First generation: eye-in-head measurement of the eye consisting of techniques such as scleral contact lens/search coil, electro-oculography
2. Second generation: photo- and video-oculography
3. Third generation: analog video-based combined pupil/corneal reflection
4. Fourth generation: digital video-based combined pupil/corneal reflection, augmented by computer vision techniques and Digital Signal Processors (DSPs)

Often the most desired type of eye tracking output (e.g., for human–computer interaction usability studies) is estimation of the projected Point Of Regard (POR) of the viewer, i.e., the (x,y) coordinates of the user's gaze on the computer display. First- and second-generation eye trackers generally do not provide this type of data. (For second-generation systems, eye movement analysis relies on off-line, frame-by-frame visual inspection of photographs or video frames and does not allow easy POR calculation.) Combined video-based pupil/corneal reflection eye trackers easily provide POR coordinates following calibration, and are today *de rigeur*. Due to the availability of fast analog-to-digital video processors, these third-generation eye trackers are capable of delivering the calculated POR in real-time. Fourth-generation eye trackers, having recently appeared on the market, make use of digital optics. Coupled with on-chip Digital Signal Processors (DSPs), eye tracking technology has significantly increased in its usability, accuracy, and speed while decreasing in cost.

The state of today's technology can best be summarized by a brief functional comparison of older and newer equipment, given in the following table. If, to those new to this book and uninitiated with eye tracking devices, the comparison in the table is

Functional eye tracker comparison.

	Legacy Systems	State-of-the-Art
Technology	Analog video	Digital video
Calibration	5- or 9-point, tracker-controlled	Any number, application-controlled
Optics	Manual focusing/thresholding	Auto-focus
Communication	Serial (polling/streaming)	TCP/IP (client/server)
Synchronization	Status byte word	API callback

not immediately suggestive, consider two use scenarios of both old and new technology presented in the next table.

Eye tracker use comparison.

Typical session with old system.	Typical session with new system.
1. Login to console.	1. Login to console.
2. Turn on eye tracking equipment.	2. Turn on eye tracking PC.
3. Turn on eye/scene monitors.	3. Run eye tracking program.
4. Turn on eye tracking PC.	4. Calibrate.
5. Run eye tracking program.	5. Run.
6. Turn on camera.	
7. Turn on illumination control.	
8. Adjust head/chin rest.	
9. Adjust pan/tilt unit.	
10. Adjust camera zoom.	
11. Adjust camera focus.	
12. Adjust pupil/corneal thresholds.	
13. Calibrate.	
14. Run.	

The disparity in the use of old and new technologies is mainly due to the use of different optics (camera). New systems tend to use an auto-focusing digital camera, e.g., embedded in a flat panel display. Although embedding within a flat panel display may restrict a user's physical position somewhat, it is generally preset to operate at a comfortable range (e.g., 50–60 cm focal distance). Unlike older systems, as long as the user sits within this distance, no chin rests and no further parameter adjustments are needed. In contrast, older devices required the use of a pan/tilt unit to position the camera, the camera to be manually focused and zoomed, and software to be set to appropriate pupil and corneal reflection detection thresholds. None of these cumbersome operations are required with newer systems.

Furthermore, one of the most important features of the new technology, especially for application development, is an individual's ability to self-calibrate. With older technology, whenever a developer wished to test a new feature, she or he had to recruit a (very patient) subject for testing. This was quite problematic. The newer systems' calibration routines are a much-needed improvement over older (third-generation) technology that significantly accelerate program and application development.

A third-generation eye tracker was used for most of the eye tracking research on which the first edition of this book was based. The availability of new technology precipitated the writing of the second edition. The second edition therefore fills several important gaps not covered previously, namely:

1. Client/server model for developing an eye tracking client application
2. Client-controlled display, calibration, data collection
3. New programming examples

Beyond updated technical descriptions of client programming, the second edition also includes what the first edition lacked: an overview of the methodology behind the use of eye trackers, that is, experimental design issues that are often needed to conduct eye tracking studies. The second edition briefly reviews experimental design decisions, offers some guidelines for incorporating eye movement metrics into a study (e.g., usability), and provides examples of case studies.

Finally, the second edition expands the third part of the book: eye tracking applications. A great deal of new and exciting eye tracking work has appeared, undoubtedly driven by the availability of new technology. In fact, there now appears to be a rather refreshing shift in the reporting of eye tracking and eye movement studies. Authors now tend to understate the "gee-whiz" factor of eye trackers and their technical machinations needed to obtain eye movement data and are now emphasizing scientific results bolstered by objective evidence provided by users' gaze and hence attention. Eye tracking finally appears to be entering into mainstream science, where the eye tracker is becoming less of a novelty and more of a tool. It is hoped that this second edition may inspire readers with the simplicity of application development now made possible by fourth-generation eye trackers and continue on the road to new applications and scientific insights.

Andrew T. Duchowski
Clemson, SC, April 2006

Preface to the First Edition

The scope of the book falls within a fairly narrow human–computer interaction domain (i.e., describing a particular input modality), however, it spans a broad range of interdisciplinary research and application topics. There are at least three domains that stand to benefit from eye tracking research: visual perception, human–computer interaction, and computer graphics. The amalgamation of these topics forms a symbiotic relationship. Graphical techniques provide a means of generating rich sets of visual stimuli ranging from 2D imagery to 3D immersive virtual worlds and research exploring visual attention and perception in turn influences the generation of artificial scenes and worlds. Applications derived from these disciplines create a powerful human–computer interaction modality, namely interaction based on knowledge of the user's gaze.

Recent advancements in eye tracking technology, specifically the availability of cheaper, faster, more accurate, and easier to use trackers, have inspired increased eye movement and eye tracking research efforts. However, although eye trackers offer a uniquely objective view of overt human visual and attentional processes, eye trackers have not yet gained widespread use beyond work conducted at various research laboratories. This lack of acceptance is due in part to two reasons: first, the use of an eye tracker in an applied experimental setting is not a widely taught subject. Hence, there is a need for a book that may help in providing training. It is not uncommon for enthusiastic purchasers of eye tracking equipment to become discouraged with their newly bought equipment when they find it difficult to set up and operate. Only a few academic departments (e.g., psychology, computer science) offer any kind of instruction in the use of eye tracking devices. Second, to exacerbate the lack of training in eye tracking methodology, even fewer sources of instruction exist for system development. Setting up an eye tracking lab and integrating the eye tracker into an available computer system for development of gaze-contingent applications is a fairly complicated endeavor, similar to the development and integration of virtual reality programs. Thus far, it appears no textbook other than this one exists providing this type of low-level information.

The goal of this book is to provide technical details for implementation of a gaze-contingent system, couched in the theoretical context of eye movements, visual perception, and visual attention. The text started out as the author's personal notes on the integration of a commercial eye tracker into a virtual reality graphics system. These technical considerations comprise the middle chapters of the book and include details of integrating a commercial eye tracker into both a 3D virtual environment, and a 2D image display application. The surrounding theoretical review chapters grew from notes developed for an interdisciplinary eye tracking methodology course offered to both undergraduates and graduates from four disciplines: psychology, marketing, industrial engineering, and computer science. An early form of these notes was presented as a short course at the Association for Computing Machinery (ACM) Special Interest Group on Graphics' SIGGRAPH conference, 23–28 July 2000, New Orleans, LA.

Overview

As of the second edition, the book is divided into four parts, presented thematically in a top-down fashion, providing first an introduction to the human visual system (Part I), then briefly surveying eye tracking systems (Part II), then discussing eye tracking methodology (Part III), and finally ending by reviewing a number of eye tracking applications (Part IV).

In the first part, "Introduction to the Human Visual System (HVS)," the book covers the concept of visual attention, mainly from a historical perspective. The first chapter focuses on the dichotomy of foveal and peripheral vision (the "what" versus the "where"). This chapter covers easily observable attentional phenomena. The next chapter covers the neurological substrate of the HVS presenting the low-level neurological elements implicated in dynamic human vision. This chapter discusses the primary dual pathways, the parvo- and magno-cellular channels, which loosely correspond to the flow of visual information permitted by the retinal fovea and periphery. Following this description of the visual "hardware", observable characteristics of human vision are summarized in the following chapter on visual perception. Here, results obtained mainly from psychophysics are summarized, distinguishing foveal and peripheral visual perception. The first part ends by discussing the mechanism responsible for shifting the fovea, namely eye movements. Having established the neurological and psychophysical context for eye movements, the following chapter on the taxonomy and models of eye movements gives the common terms for the most basic of eye movements along with a signal-analytic description of recordable eye movement waveforms.

The second part of the book, "Eye Tracking Systems," presents a brief survey of the main types of available eye tracking devices, followed by a detailed technical description of the requirements for system installation and application program development. These details are mainly applicable to video-based, corneal-reflection eye trackers, the most widely available and most affordable type of eye trackers. This part of the book offers information for the development of three general systems: one for binocular 3D eye tracking in virtual reality, one for monocular 2D eye tracking over a 2D display (e.g., a television monitor on which graphical information can be displayed), and one for binocular 2D eye tracking on the desktop. The descriptions of the first two former systems are very similar because they are based on the same kind of (older) eye tracking hardware (ISCAN in this instance). The latter system description is based on modern eye tracking technology from Tobii. Both system descriptions include notes on system calibration. This part of the book ends with a description of data collection and analysis independent of any particular eye tracking hardware.

The fourth part of the book surveys a number of interesting and challenging eye tracking applications. Applications identified in this part are drawn from psychology, human factors, marketing and advertising, human–computer interaction and collaborative systems, and computer graphics and virtual reality.

How to Read This Book

The intended audience for this book is an interdisciplinary one, aimed particularly at those interested in psychology, marketing, industrial engineering, and computer science. Indeed, this text is meant for undergraduates and graduates from these disciplines enrolled in a course dealing with eye tracking, such as the eye tracking methodology course developed by the author at Clemson University. In this course, typically all chapters are covered, but not necessarily in the order presented in the text. In such a course, the order of chapters may be as follows.

First, Part IV is presented outlining various eye tracking applications. Normally, this part should give the reader motivation for design and implementation of a semester-long eye tracking project. Coverage of this part of the book is usually supplemented by readings of research papers from various sources. For example, papers may be selected from the following conferences.

- The proceedings of the Eye Tracking Research & Applications (ETRA) conference
- The proceedings of the ACM Special Interest Group on Human–Computer Interaction (SIGCHI) conference (Human Factors in Computing)
- Transactions on Graphics, the proceedings of the annual Association for Computing Machinery (ACM) Special Interest Group on Graphics and Interactive Techniques (SIGGRAPH) conference series
- The proceedings of the Human Factors and Ergonomics Society (HFES)

To speed up development of an eye tracking application, Part II follows the presentation of Part IV, dealing in the technical details of eye tracker application development.

The types of applications that can be expected of students will depend mainly on the programming expertise represented by members of interdisciplinary student teams. For example, in the eye tracking methodology course at Clemson, teams are formed by joining computer science students with one or more of the other representatives enrolled in the class (i.e., from marketing, psychology, or industrial engineering). Although all group members decide on a project, students studying the latter subjects are mainly responsible for the design and analysis of the eventual eye tracking experiment.

Given commencement of an eye tracking application, Part III is then covered, going over experimental design. In the context of the usability measurement framework, the eye tracking methodology course advocates performance measurement, and therefore focuses on laboratory experiments and quantitative data analysis.

Part I of the text is covered last, giving students the necessary theoretical context for the eye tracking pilot study. Thus, although the book is arranged "top-down", the course proceeds "bottom-up".

The book is also suitable for researchers interested in setting up an eye tracking laboratory and/or using eye trackers for conducting experiments. Because readers with these goals may also come from diverse disciplines such as marketing, psychology, industrial engineering, and computer science, not all parts of the book may be suitable for everyone. More technically oriented readers will want to pay particular attention to the middle sections of the book which detail system installation and implementation of eye tracking application software. Readers not directly involved with such low-level details may wish to omit these sections and concentrate more on the theoretical and historical aspects given in the front sections of the book. The latter part of the book, dealing with eye tracking applications, should be suitable for all readers inasmuch as it presents examples of current eye tracking research.

Acknowledgments

This work was supported in part by a University Innovation grant (# 1-20-1906-51-4087), NASA Ames task (# NCC 2-1114), and NSF CAREER award # 9984278.

The preparation of this book has been assisted by many people, including Keith Karn, Roel Vertegaal, Dorion Liston, and Keith Rayner who provided comments on early editions of the text-in-progress. Later versions of the draft were reviewed by external reviewers to whom I express my gratitude, for their comments greatly improved the final version of the text. Special thanks go to David Wooding for his careful and thorough review of the text.

I would like to thank the team at Springer for helping me compose the text. Thanks go to Beverly Ford and Karen Borthwick for egging me on to write the text and to

Rosie Kemp and Melanie Jackson for helping me with the final stages of publication. Many thanks to Catherine Brett for her help in the creation of the second edition.

Special thanks go to Bruce McCormick, who always emphasized the importance of writing during my doctoral studies at Texas A&M University, College Station, TX. Finally, special thanks go to Corey, my wife, for patiently listening to my various ramblings on eye movements, and for being an extremely patient eye tracking subject :).

I have gained considerable pleasure and enjoyment in putting the information I've gathered and learned on paper. I hope that readers of this text derive similar pleasure in exploring vision and eye movements as I have, and they go on to implementing ever interesting and fascinating projects—have fun!

Andrew T. Duchowski
Clemson, SC, June 2002 & July 2006

Contents

Part IV Eye Tracking Applications

List of Figures

List of Tables

Introduction to the Human Visual System (HVS)

1

Visual Attention

In approaching the topic of eye tracking, we first have to consider the motivation for recording human eye movements. That is, why is eye tracking important? Simply put, we move our eyes to bring a particular portion of the visible field of view into high resolution so that we may see in fine detail whatever is at the central direction of gaze. Most often we also divert our attention to that point so that we can focus our concentration (if only for a very brief moment) on the object or region of interest. Thus, we may presume that if we can track someone's eye movements, we can follow along the path of attention deployed by the observer. This may give us some insight into what the observer found interesting, that is, what drew their attention, and perhaps even provide a clue as to how that person perceived whatever scene she or he was viewing.

By examining attention and the neural mechanisms involved in visual attention, the first two chapters of this book present motivation for the study of eye movements from two perspectives: a psychological viewpoint examining attentional behavior and its history of study (presented briefly in this chapter); and a physiological perspective on the neural mechanisms responsible for driving attentional behavior (covered in the next chapter). In sum, both introductory chapters establish the psychological and physiological basis for the movements of the eyes.

To begin formulating an understanding of an observer's attentional processes, it is instructive to first establish a rudimentary or at least intuitive sense of what attention is, and whether the movement of the eyes does in fact disclose anything about the inner cognitive process known as visual attention.

Visual attention has been studied for over a hundred years. A good qualitative definition of visual attention was given by the psychologist William James:

> Everyone knows what attention is. It is the taking possession by the mind, in clear and vivid form, of one out of what seem several simultaneously possible objects or trains of thought. Focalization, concentration, of consciousness

are of its essence. It implies withdrawal from some things in order to deal effectively with others...

When the things are apprehended by the *senses*, the number of them that can be attended to at once is small, '*Pluribus intentus, minor est ad singula sensus.*'

—W. James (1981)

The Latin phrase used above by James roughly translates to "*Many filtered into few for perception.*" The faculty implied as the filter is attention.

Humans are finite beings that cannot attend to all things at once. In general, attention is used to focus our mental capacities on selections of the sensory input so that the mind can successfully process the stimulus of interest. Our capacity for information processing is limited. The brain processes sensory input by concentrating on specific components of the entire sensory realm so that interesting sights, sounds, smells, and the like, may be examined with greater attention to detail than peripheral stimuli. This is particularly true of vision. Visual scene inspection is performed *minutatim*, not *in toto*. That is, human vision is a piecemeal process relying on the perceptual integration of small regions to construct a coherent representation of the whole.

In this chapter, attention is recounted from a historical perspective following the narrative found in Heijden (1992). The discussion focuses on attentional mechanisms involved in vision, with emphasis on two main components of visual attention, namely the "what" and the "where".

1.1 Visual Attention: A Historical Review

The phenomenon of visual attention has been studied for over a century. Early studies of attention were technologically limited to simple ocular observations and oftentimes to introspection. Since then the field has grown to an interdisciplinary subject involving the disciplines of psychophysics, cognitive neuroscience, and computer science, to name three. This section presents a qualitative historical background of visual attention.

1.1.1 Von Helmholtz's "Where"

At the second half of the 19th century, Von Helmholtz (1925) posited visual attention as an essential mechanism of visual perception. In his *Treatise on Physiological Optics*, he notes, "We let our eyes roam continually over the visual field, because that is the only way we can see as distinctly as possible all the individual parts of the field in turn." Noting that attention is concerned with a small region of space,

Von Helmholtz observed visual attention's natural tendency to wander to new things. He also remarked that attention can be controlled by a conscious and voluntary effort, allowing attention to peripheral objects without making eye movements to that object. Von Helmholtz was mainly concerned with eye movements to spatial locations, or the "where" of visual attention. In essence, although visual attention can be consciously directed to peripheral objects, eye movements reflect the will to inspect these objects in fine detail. In this sense, eye movements provide evidence of overt visual attention.

1.1.2 James' "What"

In contrast to Von Helmholtz's ideas, James (1981) believed attention to be a more internally covert mechanism akin to imagination, anticipation, or in general, thought. James defined attention mainly in terms of the "what", or the identity, meaning, or expectation associated with the focus of attention. James favored the active and voluntary aspects of attention although he also recognized its passive, reflexive, non-voluntary and effortless qualities.

Both views of attention, which are not mutually exclusive, bear significantly on contemporary concepts of visual attention. The "what" and "where" of attention roughly correspond to foveal (James) and parafoveal (Von Helmholtz) aspects of visual attention, respectively. This dichotomous view of vision is particularly relevant to a bottom-up or feature-driven explanation of visual attention. That is, when considering an image stimulus, we may consider certain regions in the image that will attract one's attention. These regions may initially be perceived parafoveally, in a sense requesting further detailed inspection through foveal vision. In this sense, peripherally located image features may drive attention in terms of "where" to look next, so that we may identify "what" detail is present at those locations.

The dual "what" and "where" feature-driven view of vision is a useful preliminary metaphor for visual attention, and indeed it has formed the basis for creating computational models of visual attention, which typically simulate so-called low-level, or bottom-up visual characteristics. However, this view of attention is rather simplistic. It must be stressed that a complete model of visual attention involves high-level visual and cognitive functions. That is, visual attention cannot simply be explained through the sole consideration of visual features. There are higher-level intentional factors involved (e.g., related to possibly voluntary, preconceived cognitive factors that drive attention).

1.1.3 Gibson's "How"

In the 1940s Gibson (1941) proposed a third factor of visual attention centered on intention. Gibson's proposition dealt with a viewer's advance preparation as to whether to react and if so, how, and with what class of responses. This component of attention explained the ability to vary the intention to react while keeping the expectation of the stimulus object fixed, and conversely, the ability to vary the expectation of the stimulus object while keeping the intention to react fixed. Experiments involving ambiguous stimuli typically evoke these reactions. For example, if the viewer is made to expect words describing animals, then the misprint "sael" will be read as "seal". Changing the expectation to words describing ships or boats invokes the perception of "sail". The reactive nature of Gibson's variant of attention specifies the "what to do", or "how to react" behavior based on the viewer's preconceptions or attitude. This variant of visual attention is particularly relevant to the design of experiments. It is important to consider the viewer's perceptual expectation of the stimulus, as (possibly) influenced by the experimenter's instructions.

1.1.4 Broadbent's "Selective Filter"

Attention, in one sense, is seen as a "selective filter" responsible for regulating sensory information to sensory channels of limited capacity. In the 1950s, Broadbent (1958) performed auditory experiments designed to demonstrate the selective nature of auditory attention. The experiments presented a listener with information arriving simultaneously from two different channels, e.g., the spoken numerals $\{7,2,3\}$ to the left ear, $\{9,4,5\}$ to the right. Broadbent reported listeners' reproductions of either $\{7,2,3,9,4,5\}$, or $\{9,4,5,7,2,3\}$, with no interwoven (alternating channel) responses. Broadbent concluded that information enters in parallel but is then selectively filtered to sensory channels.

1.1.5 Deutsch and Deutsch's "Importance Weightings"

In contrast to the notion of a selective filter, Deutsch and Deutsch (1963) proposed that all sensory messages are perceptually analyzed at the highest level, precluding a need for a selective filter. Deutsch and Deutsch rejected the selective filter and limited capacity system theory of attention; they reasoned that the filter would need to be at least as complex as the limited capacity system itself. Instead, they proposed the existence of central structures with preset "importance weightings" that determined selection. Deutsch and Deutsch argued that it is not attention as such but the weightings of importance that have a causal role in attention. That is, attentional effects are a result of importance, or relevance, interacting with the information.

It is interesting to note that Broadbent's selective filter generally corresponds to Von Helmholtz's "where", whereas Deutsch and Deutsch's importance weightings correspond to James' expectation, or the "what". These seemingly opposing ideas were incorporated into a unified theory of attention by Anne Treisman in the 1960s (although not fully recognized until 1971). Treisman brought together the attentional models of Broadbent and Deutsch and Deutsch by specifying two components of attention: the attenuation filter followed by later (central) structures referred to as "dictionary units". The attenuation filter is similar to Broadbent's selective filter in that its function is selection of sensory messages. Unlike the selective filter, it does

not completely block unwanted messages, but only attenuates them. The later stage
dictionary units then process weakened and unweakened messages. These units contain variable thresholds tuned to importance, relevance, and context. Treisman thus
brought together the complementary models of attentional unit or selective filter (the
"where"), and expectation (the "what").

Up to this point, even though Treisman provided a convincing theory of visual attention, a key problem remained, referred to as the scene integration problem. The
scene integration problem poses the following question: even though we may view
the visual scene through something like a selective filter, which is limited in its scope,
how is it that we can piece together in our minds a fairly coherent scene of the entire
visual field? For example, when looking at a group of people in a room such as in a
classroom or at a party, even though it is impossible to gain a detailed view of everyone's face at the same time, nevertheless it is possible to assemble a mental picture
of where people are located. Our brains are capable of putting together this mental
picture even though the selective filter of vision prevents us from physically doing
so in one glance. Another well-known example of the scene integration problem is
the Kanizsa (1976) illusion, exemplified in Figure 1.1, named after the person who
invented it. Inspecting Figure 1.1, you will see the edges of a triangle, even though
the triangle is defined only by the notches in the disks. How this triangle is integrated
by the brain is not yet fully understood. That is, although it is known that the scene
is inspected piecemeal as evidenced by the movement of the eyes, it is not clear how
the "big picture" is assembled, or integrated, in the mind. This is the crux of the
scene integration problem. In one view, offered by the Gestalt psychologists, it is
hypothesized that recognition of the entire scene is performed by a parallel one-step
process. To examine this hypothesis, a visualization of how a person views such an
image (or any other) is particularly helpful. This is the motivation for recording and
visualizing a viewer's eye movements. Even though the investigation of eye movements dates back to 1907 (Dodge, 1907),[1] a clear depiction of eye movements would
not be available until 1967 (see Chapter 5 for a survey of eye tracking techniques).
This early eye movement visualization, discussed below, shows the importance of
eye movement recording not only for its expressive power of depicting one's visual
scanning characteristics, but also for its influence on theories of visual attention and
perception.

1.1.6 Yarbus and Noton and Stark's "Scanpaths"

Early diagrammatic depictions of recorded eye movements helped cast doubt on the
Gestalt hypothesis that recognition is a parallel one-step process. The Gestalt view
of recognition is a holistic one suggesting that vision relies to a great extent on the
tendency to group objects. Although well-known visual illusions exist to support this
view (e.g., subjective contours of the Kanizsa figure; see Figure 1.1 above Kanizsa
(1976)), early eye movement recordings showed that visual recognition is at least

---

[1] As cited in Gregory (1990).

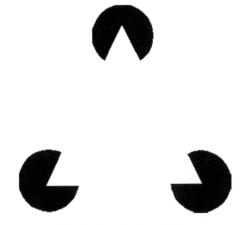

Fig. 1.1. The Kanizsa illusion.

partially serial in nature.

Yarbus (1967) measured subjects' eye movements over an image after giving subjects specific questions related to the image. Such a picture is shown in Figure 1.2. Questions posed to subjects included a range of queries specific to the situation, e.g., are the people in the image related, what are they wearing, what will they have to eat, and so on. The eye movements Yarbus recorded demonstrated sequential viewing patterns over particular regions in the image.

Noton and Stark (1971a, 1971b) performed their own eye movement measurements over images and coined the observed patterns "scanpaths". Their work extended Yarbus' results by showing that even without leading questions subjects tend to fixate identifiable regions of interest, or "informative details". Furthermore, scanpaths showed that the order of eye movements over these regions is quite variable. That is, given a picture of a square, subjects will fixate on the corners, although the order in which the corners are viewed differs from viewer to viewer and even differs between consecutive observations made by the same individual.

In contrast to the Gestalt view, Yarbus' and Noton and Stark's work suggests that a coherent picture of the visual field is constructed piecemeal through the assembly of serially viewed regions of interest. Noton and Stark's results support James' "what" of visual attention. With respect to eye movements, the "what" corresponds to regions of interest selectively filtered by foveal vision for detailed processing.

1.1.7 Posner's "Spotlight"

Contrary to the serial "what" of visual attention, the orienting, or the "where", is performed in parallel (Posner et al., 1980). Posner et al. suggested an attentional mechanism able to move about the scene in a manner similar to a "spotlight." The

In each of the traces, the subject was asked to: Trace 1, examine the picture at will; Trace 2, estimate the economic level of the people; Trace 3, estimate the people's ages; Trace 4, guess what the people were doing before the arrival of the visitor; Trace 5, remember the people's clothing; Trace 6, remember the people's (and objects') position in the room; Trace 7, estimate the time since the guest's last visit.

Fig. 1.2. Yarbus' early eye movement recordings. Reprinted from Yarbus (1967) with permission © 1967 Plenum Press.

spotlight, being limited in its spatial extent, seems to fit well with Noton and Stark's empirical identification of foveal regions of interest. Posner et al., however, dissociate the spotlight from foveal vision and consider the spotlight an attentional mechanism independent of eye movements. Posner et al. identified two aspects of visual attention: the orienting and the detecting of attention. Orienting may be an entirely central (covert or mental) aspect of attention, whereas detecting is context-sensitive, requiring contact between the attentional beam and the input signal. The orienting of attention is not always dependent on the movement of the eyes; that is, it is possible to attend to an object while maintaining gaze elsewhere. According to Posner et al., orientation of attention must be done in parallel and must precede detection.

The dissociation of attention from foveal vision is an important point. In terms of the "what" and the "where", it seems likely that the "what" relates to serial foveal vision. The "where", on the other hand, is a parallel process performed parafoveally, or peripherally, which dictates the next focus of attention.

1.1.8 Treisman's "Glue"

Posner et al. and Noton and Stark advanced the theory of visual attention along similar lines forged by Von Helmholtz and James (and then Broadbent and Deutsch and Deutsch). Treisman once again brought these concepts together in the feature integration theory of visual attention (Treisman & Gelade, 1980; Treisman, 1986). In essence, attention provides the "glue" that integrates the separated features in a particular location so that the conjunction (i.e., the object) is perceived as a unified whole. Attention selects features from a master map of locations showing *where* all the feature boundaries are located, but not *what* those features are. That is, the master map specifies where things are, but not what they are. The feature map also encodes simple and useful properties of the scene such as color, orientation, size, and stereo distance. Feature Integration Theory, or FIT, is a particularly important theory of visual attention and visual search. Eye tracking is often a significant experimental component used to test FIT. Feature integration theory, treated as an eye tracking application, is discussed in more detail in Chapter 17.

1.1.9 Kosslyn's "Window"

Recently, Kosslyn (1994) proposed a refined model of visual attention. Kosslyn describes attention as a selective aspect of perceptual processing, and proposes an attentional "window" responsible for selecting patterns in the "visual buffer". The window is needed because there is more information in the visual buffer than can be passed downstream, and hence the transmission capacity must be selectively allocated. That is, some information can be passed along, but other information must be filtered out. This notion is similar to Broadbent's selective filter and Treisman's attenuation filter. The novelty of the attentional window is its ability to be adjusted incrementally; i.e., the window is scalable. Another interesting distinction of Kosslyn's model is the hypothesis of a redundant stimulus-based attention-shifting subsystem (e.g., a type of context-sensitive spotlight) in mental imagery. Mental imagery involves the formation of mental maps of objects, or of the environment in general. It is defined as "...the mental invention or recreation of an experience that in at least some respects resembles the experience of actually perceiving an object or an event, either in conjunction with, or in the absence of, direct sensory stimulation" (Finke, 1989). It is interesting to note that the eyes move during sleep (known as Rapid Eye Movement or REM sleep). Whether this is a manifestation of the use of an internal attentional window during sleep is not known.

1.2 Visual Attention and Eye Movements

Considering visual attention in terms of the "what" and "where", we would expect that eye movements work in a way that supports the dual attentive hypothesis. That is, vision might behave in a cyclical process composed of the following steps.

1. Given a stimulus, such as an image, the entire scene is first seen mostly in parallel through peripheral vision and thus mostly at low resolution. At this stage, interesting features may "pop out" in the field of view, in a sense engaging or directing attention to their location for further detailed inspection.
2. Attention is thus turned off or disengaged from the foveal location and the eyes are quickly repositioned to the first region that attracted attention.
3. Once the eyes complete their movement, the fovea is now directed at the region of interest, and attention is now engaged to perceive the feature under inspection at high resolution.

This is a bottom-up model or concept of visual attention. If the model is accurate, one would expect to find regions in the brain that correspond in their function to attentional mechanisms. This issue is further investigated in Chapter 2.

The bottom-up model is at least correct in the sense that it can be said to be a component of natural human vision. In fact, the bottom-up model forms a powerful basis for computational models of visual search. Examples of such models are presented later in the text (see Chapter 17). The bottom-up view of visual attention is, however, incomplete. There are several key points that are not addressed. Consider these questions:

1. Assuming it is only the visual stimulus (e.g., image features) that drives attention, exactly what types of features attract attention?
2. If the visual stimulus were solely responsible for attracting attention, would we ever need the capability of making voluntary eye movements?
3. What is the link between attention and eye movements? Is attention always associated with the foveally viewed portion of the visual scene?

To gain insight into the first question, we must examine how our physical visual mechanism (our eyes and brain) responds to visual stimulus. To attempt to validate a model of visual attention, we would need to be able to justify the model by identifying regions in the brain that are responsible for carrying out the functionality proposed by the model. For example, we would expect to find regions in the brain that engage and disengage attention as well as those responsible for controlling (i.e., programming, initiating, and terminating) the movements of the eyes. Furthermore, there must be regions in the brain that are responsible for responding to and interpreting the visual stimuli that are captured by the eyes. As shown in the following chapters, the Human Visual System (HVS) responds strongly to some types of stimuli (e.g., edges), and weakly to others (e.g., homogeneous areas). The following chapters show that this response can be predicted to a certain extent by examining the physiology of the HVS. In later chapters we also show that the human visual response can be measured through a branch of psychology known as psychophysics. That is, through psychophysics, we can fairly well measure the perceptive power of the human visual system.

The bottom-up model of visual attention does not adequately offer answers to the second question because it is limited to mostly bottom-up, or feature-driven aspects of attention. The answer to the second question becomes clearer if we consider a more complete picture of attention involving higher-level cognitive functions. That is, a complete theory of visual attention should also involve those cognitive processes that describe our voluntary intent to attend to something, e.g., some portion of the scene. This is a key point that was briefly introduced following the summary of Gibson's work, and which is evident in Yarbus' early scanpaths. It is important to reiterate that Yarbus' work demonstrated scanpaths which differed with observers' expectations; that is, scanpath characteristics such as their order of progression can be *task-dependent*. Based on what they are looking for, people will view a picture differently. A complete model or theory of visual attention is beyond the scope of this book, but see Chapter 17 for further insight into theories of visual search, and also for examples of the application of eye trackers to study this question.

Considering the third question opens up a classical problem in eye tracking studies. Because attention is composed of both low-level and high-level functions (one can loosely think of involuntary and voluntary attention, respectively), as Posner and others have observed, humans can voluntarily dissociate attention from the foveal direction of gaze. In fact, astronomers do this regularly to detect faint constellations with the naked eye by looking "off the fovea." Because the periphery is much more sensitive to dim stimulus, faint stars are much more easily seen out of the "corner" of one's eye than when they are viewed centrally. Thus the high-level component of vision may be thought of as a covert component, or a component which is not easily detectable by external observation. This is a well-known problem for eye tracking researchers. An eye tracker can only track the overt movements of the eyes, however, it cannot track the covert movement of visual attention. Thus, in all eye tracking work, a tacit but very important assumption is usually accepted: we assume that attention is linked to foveal gaze direction, but we acknowledge that it may not always be so.

1.3 Summary and Further Reading

A historical account of attention is a prerequisite to forming an intuitive impression of the selective nature of perception. For an excellent historical account of selective visual attention, see Heijden (1992). An earlier and very readable introduction to visual processes is a small paperback by Gregory (1990). For a more neurophysiological perspective, see Kosslyn (1994). Another good text describing early attentional vision is Papathomas et al. (1995).

The singular idioms describing the selective nature of attention are the "what" and the "where". The "where" of visual attention corresponds to the visual selection of specific regions of interest from the entire visual field for detailed inspection. Notably, this selection is often carried out through the aid of peripheral vision. The "what" of visual attention corresponds to the detailed inspection of the spatial region through a perceptual channel limited in spatial extent. The attentional "what" and "where" duality is relevant to eye tracking studies because scanpaths show the temporal progression of the observer's foveal direction of gaze and therefore depict the observer's instantaneous overt localization of visual attention.

From investigation of visual search, the consensus view is that a parallel pre-attentive stage acknowledges the presence of four basic features: color, size, orientation, and presence and/or direction of motion and that features likely to attract attention include edges and corners, but not plain surfaces (see Chapter 17). There is some doubt, however, whether human visual search can be described as an integration of independently processed features (Van Orden & DiVita, 1993). Van Orden and DiVita suggest that "...any theory on visual attention must address the fundamental properties of early visual mechanisms." To attempt to quantify the visual system's processing capacity, the neural substrate of the human visual system is examined in the following chapter which surveys the relevant neurological literature.

2

Neurological Substrate of the HVS

Considerable information may be gleaned from the vast neuroscientific literature regarding the functionality (and limitations) of the Human Visual System (HVS). It is often possible to qualitatively predict observed psychophysical results by studying the underlying visual "hardware." For example, visual spatial acuity may be roughly estimated from knowledge of the distribution of retinal photoreceptors. Other characteristics of human vision may also be estimated from the neural organization of deeper brain structures.

Neurophysiological and psychophysical literature on the human visual system suggests the field of view is inspected *minutatim* through brief fixations over small regions of interest. This allows perception of detail through the fovea. Central foveal vision subtends 1–5° (visual angle) allowing fine scrutiny of only a small portion of the entire visual field, for example only 3% of the size of a large (21 in.) computer monitor (seen at ~60 cm viewing distance). Approximately 90% of viewing time is spent in fixations. When visual attention is directed to a new area, fast eye movements (saccades) reposition the fovea. The dynamics of visual attention probably evolved in harmony with (or perhaps in response to) the perceptual limitations imposed by the neurological substrate of the visual system.

The brain is composed of numerous regions classified by their function (Zeki, 1993). A simplified representation of brain regions is shown in Figure 2.1, with lobe designations stylized in Figure 2.2. The human visual system is functionally described by the connections between retinal and brain regions, known as visual pathways. Pathways joining multiple brain areas involved in common visual functions are referred to as streams. Figure 2.1 highlights regions and pathways relevant to selective visual attention. For clarity, many connections are omitted. Of particular importance to dynamic visual perception and eye movements are the following neural regions, summarized in terms of relevance to attention.

- SC (Superior Colliculus): involved in programming eye movements and contributes to eye movement target selection for both saccades and smooth pursuits (possibly in concert with the Frontal Eye Fields (FEF) and area Lateral IntraParietal (LIP)); also remaps auditory space into visual coordinates (presumably for target foveation); with input of motion signals from area MT (see below), the SC is involved in pursuit target selection as well as saccade target selection.

- Area V1 (primary visual cortex): detection of range of stimuli, e.g., principally orientation selection and possibly to a lesser extent color; cellular blob regions (double-opponent color cells) respond to color variations and project to areas V2 and V4 (Livingstone & Hubel, 1988).
- Areas V2, V3, V3A, V4, MT: form, color, and motion processing.
- Area V5/MT (Middle Temporal) and MST (Middle Superior Temporal): furnish large projections to Pons; hence possibly involved in smooth pursuit movements; involved in motion processing: area MT also projects to the colliculus, providing it with motion signals from the entire visual field.
- Area LIP (Lateral Intra Parietal): contains receptive fields that are corrected (reset) before execution of saccadic eye movements.
- PPC (Posterior Parietal Complex): involved in fixations.

Connections made to these areas from area V1 can be generally divided into two streams: the dorsal and ventral streams. Loosely, their functional description can be summarized as

- Dorsal stream: sensorimotor (motion, location) processing (e.g., the attentional "where")
- Ventral stream: cognitive processing (e.g., the attentional "what")

In general attentional terms, the three main neural regions implicated in eye movement programming and their functions are (Palmer, 1999):

- Posterior Parietal Complex: disengages attention,
- SC: relocates attention,
- Pulvinar: engages, or enhances, attention.

In a very simplified view of the brain, it is possible to identify the neural mechanisms involved in visual attention and responsible for the generation of eye movements. First, by examining the structure of the eye, it becomes clear why only the central or foveal region of vision can be perceived at high resolution. Second, signals from foveal and peripheral regions of the eye's retina can be roughly traced along pathways in the brain showing how the brain may process the visual scene. Third, regions in the brain can be identified which are thought to be involved in moving the eyes so that the scene can be examined piecemeal. In this simplified view of the brain, one can in a sense obtain a complete picture of an "attentional feedback loop," which creates the attentional cycles of disengaging attention, shifting of attention and (usually) the eyes, and for processing the region of interest currently being attended to, re-engaging attention and brain regions.

The neural substrate of the human visual system is examined in this chapter from the intuitive attentional perspective given above. The human neural hardware responsible

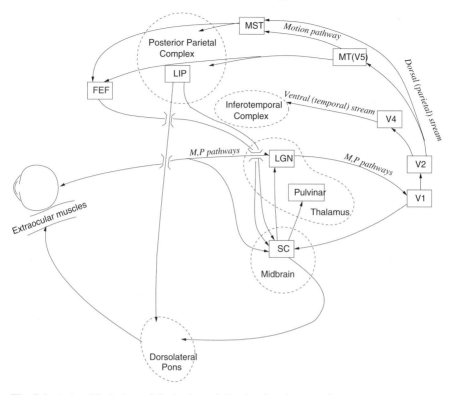

Fig. 2.1. A simplified view of the brain and the visual pathways relevant to eye movements and attention.

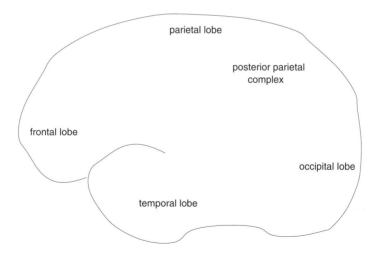

Fig. 2.2. Stylized classification of cortical lobes.

for visual processing is presented in order roughly following the direction of light and hence information entering the brain. That is, the discussion is presented "front-to-back" starting with a description of the eye and ending with a summary of the visual cortex located at the back of the brain. Emphasis is placed on differentiating the processing capability of foveal and peripheral vision, i.e., the simplified "what" and "where" of visual attention, respectively. However, the reader must be cautioned against underestimating the complexity of the visual system as presented in this text. The apparent "what" and "where" dual pathways are most probably not independent functional channels. There is a good deal of interconnection and "crosstalk" between these and other related visual centers which deems the dichotomous analysis overly simplistic. Nevertheless, there is a great deal of valuable information to be found in the neurological literature as human vision is undoubtedly the most studied human sense.

2.1 The Eye

Often called "the world's worst camera," the eye, shown in Figure 2.3, suffers from numerous optical imperfections, for example,

- Spherical aberrations: prismatic effect of peripheral parts of the lens
- Chromatic aberrations: shorter wavelengths (blue) refracted more than longer wavelengths (red)
- Curvature of field: a planar object gives rise to a curved image

However, the eye is also endowed with various mechanisms that reduce degradive effects, e.g.,

- To reduce spherical aberration, the iris acts as a stop, limiting peripheral entry of light rays,
- To overcome chromatic aberration, the eye is typically focused to produce sharp images of intermediate wavelengths,
- To match the effects of curvature of field, the retina is curved compensating for this effect.

The eye is schematically shown in Figure 2.3.

2.2 The Retina

At the rear interior surface of the eye, the retina contains receptors sensitive to light (photoreceptors) which constitute the first stage of visual perception. Photoreceptors can effectively be thought of as "transducers" converting light energy to electrical impulses (neural signals). Neural signals originating at these receptors lead to deeper visual centers in the brain. Photoreceptors are functionally classified into rods and cones. Rods are sensitive to dim and achromatic light (night vision), whereas cones

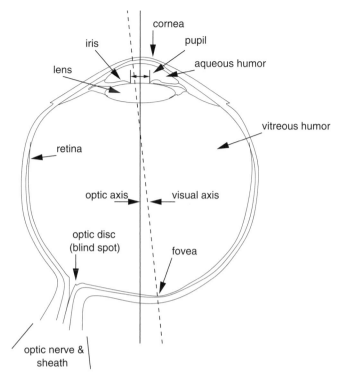

Fig. 2.3. The eye. Adapted from Visual Perception, 1st edition, by Cornsweet (1970) © 1970. Reprinted with permission of Wadsworth, a division of Thomson Learning: <www.thomsonrights.com>.

respond to brighter chromatic light (daylight vision). The retina contains approximately 120 million rods and 7 million cones.

The retina is composed of multiple layers of different cell types (De Valois & De Valois, 1988). Surprisingly, the "inverted" retina is constructed in such a way that photoreceptors are found at the bottom layer. This construction is somewhat counterintuitive inasmuch as rods and cones are farthest away from incoming light, buried beneath a layer of cells. The retina resembles a three-layer cell sandwich, with connection bundles between each layer. These connectional layers are called plexiform or synaptic layers. The retinogeniculate organization is schematically depicted in Figure 2.4. The outermost layer (w.r.t. incoming light) is the outer nuclear layer which contains the photoreceptor (rod/cone) cells. The first connectional layer is the outer plexiform layer which houses connections between receptor and bipolar nuclei. The next outer layer of cells is the inner nuclear layer containing bipolar (amacrine, bipolar, horizontal) cells. The next plexiform layer is the inner plexiform layer where

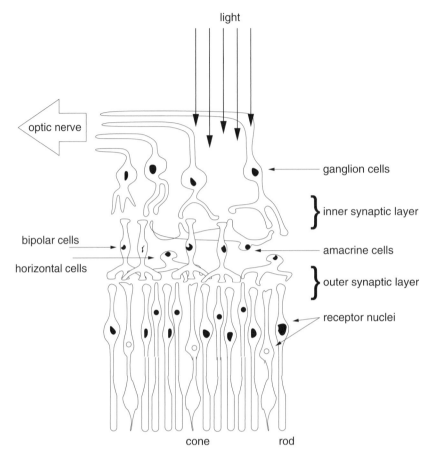

Fig. 2.4. Schematic diagram of the neural interconnections among receptors and bipolar, ganglion, horizontal, and amacrine cells. Adapted from Dowling and Boycott (1966) with permission © 1966 The Royal Society (London).

connections between inner nuclei cells and ganglion cells are formed. The top layer, or the ganglion layer, is composed of ganglion cells.

The fovea's photoreceptors are special types of neurons, the nervous system's basic elements (see Figure 2.5). Retinal rods and cones are specific types of dendrites. In general, individual neurons can connect to as many as 10,000 other neurons. Comprised of such interconnected building blocks, as a whole, the nervous system behaves as a large neural circuit. Certain neurons (e.g., ganglion cells) resemble a "digital gate," sending a signal (firing) when the cell's activation level exceeds a threshold. The myelin sheath is an axonal cover providing insulation which speeds up conduction of impulses. Unmyelinated axons of the ganglion cells converge to the optic disk (an opaque myelin sheath would block light). Axons are myelinated at

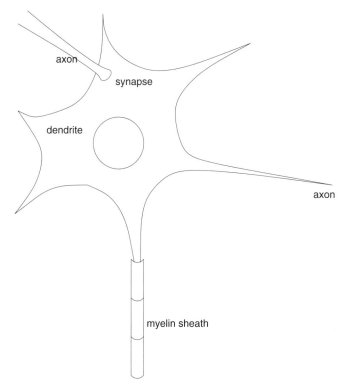

Fig. 2.5. Schematic of the neuron. From Brain, Mind, and Behavior by Floyd E. Bloom and Arlyne Lazerson © 1985, 1988, 2001 by Educational Broadcasting Corporation. Used with the permission of Worth Publishers.

the optic disk, and connect to the Lateral Geniculate Nuclei (LGN) and the Superior Colliculus (SC).

2.2.1 The Outer Layer

Rods and cones of the outer retinal layer respond to incoming light. A simplified account of the function of these cells is that rods provide monochromatic scotopic (night) vision, and cones provide trichromatic photopic (day) vision. Both types of cells are partially sensitive to mesopic (twilight) light levels.

2.2.2 The Inner Nuclear Layer

Outer receptor cells are laterally connected to the horizontal cells. In the fovea, each horizontal cell is connected to about 6 cones, and in the periphery to about 30–40 cones. Centrally, the cone bipolar cells contact one cone directly, and several cones indirectly through horizontal or receptor–receptor coupling. Peripherally, cone bipolar cells directly contact several cones. The number of receptors increases eccentrically. The rod bipolar cells contact a considerably larger number of receptors than cone bipolars. There are two main types of bipolar cells: ones that depolarize to increments of light (+), and others that depolarize to decrements of light (−). The signal profile (cross-section) of bipolar receptive fields is a "Mexican Hat," or center-surround, with an on-center, or off-center signature.

2.2.3 The Ganglion Layer

In a naive view of the human visual system, it is possible to inaccurately think of the retina (and thus the HVS as a whole) acting in a manner similar to that of a camera. Although it is true that light enters the eye and is projected through the lens onto the retina, the camera analogy is only accurate up to this point. In the retina, ganglion cells form an "active contrast-enhancing system," not a cameralike plate. Centrally, ganglion cells directly contact one bipolar. Peripherally, ganglion cells directly contact several bipolars. Thus the retinal "camera" is not composed of individual "pixels." Rather, unlike isolated pixels, the retinal photoreceptors (rods and cones in the base layer) form rich interconnections beyond the retinal outer layer. With about 120 million rods and cones and only about 1 million ganglion cells eventually innervating at the LGN, there is considerable convergence of photoreceptor output. That is, the signals of many (on the order of about 100) photoreceptors are combined to produce one type of signal. This interconnecting arrangement is described in terms of receptive fields, and this arrangement functions quite differently from a camera.

Ganglion cells are distinguished by their morphological and functional characteristics. Morphologically, there are two types of ganglion cells, the α and β cells. Approximately 10% of retinal ganglion cells are α cells possessing large cell bodies and dendrites, and about 80% are β cells with small bodies and dendrites (Lund et al., 1995). The α cells project to the magnocellular (M-) layers of the LGN and the β cells project to the parvocellular (P-) layers. A third channel of input relays through narrow, cell-sparse laminae between the main M- and P-layers of the LGN. Its origin in the retina is not yet known. Functionally, ganglion cells fall into three classes, the X, Y, and W cells (De Valois & De Valois, 1988; Kaplan, 1991). X cells respond to sustained stimulus, location and fine detail, and innervate along both M- and P-projections. Y cells innervate only along the M-projection, and respond to transient stimulus, coarse features and motion. W cells respond to coarse features, and motion, and project to the Superior Colliculus.

The receptive fields of ganglion cells are similar to those of bipolar cells (center-surround, on-center, off-center). Center-on and center-off receptive fields are depicted in Figure 2.6. Plus signs (+) denote illumination stimulus, minus signs (−) denote lack of stimulus. The vertical bars below each receptive field depict the firing response of the receptive field. This signal characteristic (series of "ticks") is usually

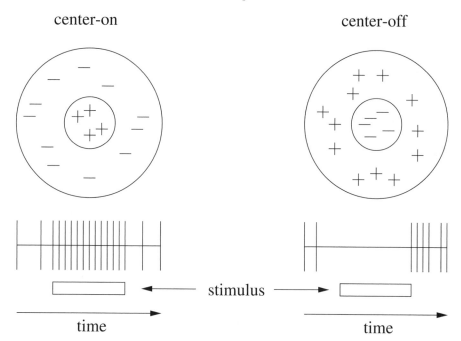

Fig. 2.6. Schematic of receptive fields.

obtained by inserting an electrode into the brain. The signal profile of receptive fields resembles the "Mexican hat" operator, often used in image processing.

2.3 The Optic Tract and M/P Visual Channels

Some (but not all) neural signals are transmitted from the retina to the occipital (visual) cortex through the optic tract, crossing in the optic chiasm, making connections to the LGN along the way. The physiology of the optic tract is often described functionally in terms of visual pathways, with reference to specific cells (e.g., ganglion cells). It is interesting to note the decussation (crossing) of the fibers from the nasal half of the retina at the optic chiasm, i.e., nasal retinal signals cross, temporal signals do not.

M and P ganglion cells in the retina connect to M and P channels, respectively. Along the optic pathways, the superior colliculus and the lateral geniculate nucleus are of particular importance. The SC is involved in programming eye movements and also remaps auditory space into visual coordinates. As shown in Figure 2.1, some neural signals along the optic tract project to the SC. The SC is thought to be responsible for directing the eyes to a new region of interest for subsequent detailed visual inspection. Like other regions in the thalamus serving similar functions, the LGN is a

crossover point, or relay station, for α and β ganglion cells. The physiological organization of the LGN, with respect to nervations of these cells, produces a visual field
topography of great clinical importance. Here, the magnocellular and the parvocellular ganglionic projections are clearly visible (under microscope), forming junctions
within two distinct layers of the LGN, correspondingly termed the M- and P-layers.
Thalamic axons from the M- and P-layers of the LGN terminate in area V1 (the primary visual center) of the striate cortex.

The functional characteristics of ganglionic projections to the LGN and the corresponding magno- and parvocellular pathways are summarized in Table 2.1. The
parvocellular pathway in general responds to signals possessing the following
attributes: high contrast (the parvocellular pathway is less sensitive to luminance),
chromaticity, low temporal frequency, and high spatial frequency (due to the small
receptive fields). Conversely, the magnocellular pathway can be characterized by
sensitivity to the following signals: low contrast (the magnocellular pathway is more
sensitive to luminance), achromaticity, moderate-to-high temporal frequency (e.g.,
sudden onset stimuli), and low spatial frequency (due to the large receptive fields).
Zeki (1993) suggests the existence of four functional pathways defined by the M and
P channels: motion, dynamic form, color, and form (size and shape). It is thought
that fibers reaching the superior colliculus represent retinal receptive fields in rod-
rich peripheral zones, whereas the fibers reaching the LGN represent cone-rich areas
of high acuity (Bloom & Lazerson, 1988). It seems likely that, in a general sense, the
M ganglion cells correspond to rods, mainly found in the periphery, and the P cells
correspond to cones, which are chromatic cells concentrated mainly in the foveal
region.

Table 2.1. Functional characteristics of ganglionic projections.

Characteristics	Magnocellular	Parvocellular
Ganglion size	Large	Small
Transmission time	Fast	Slow
Receptive fields	Large	Small
Sensitivity to small objects	Poor	Good
Sensitivity to change in light levels	Large	Small
Sensitivity to contrast	Low	High
Sensitivity to motion	High	Low
Color discrimination	No	Yes

2.4 The Occipital Cortex and Beyond

Thalamic axons from the M- and P-layers of the LGN terminate mainly in the lower
and upper halves (β, α divisions, respectively) of layer 4C in middle depth of area

V1 (Lund et al., 1995). Cell receptive field size and contrast sensitivity signatures are distinctly different in the M- and P- inputs of the LGN, and vary continuously through the depth of layer 4C. Unlike the center-surround receptive fields of retinal ganglion and LGN cells, cortical cells respond to orientation-specific stimulus (Hubel, 1988). Cortical cells are distinguished by two classes: simple and complex.

In area V1, the size of a simple cell's receptive field depends on its relative retinal position. The smallest fields are in and near the fovea, with sizes of about $1/4 \times 1/4$ degree. This is about the size of the smallest diameters of the smallest receptive field centers of retinal ganglion or LGN cells. In the far periphery, simple cell receptive field sizes are about 1×1 degree. The relationship between small foveal receptive fields and large peripheral receptive fields is maintained about everywhere along the visual pathway.

Simple cells fire only when a line or edge of preferred orientation falls within a particular location of the cell's receptive field. Complex cells fire wherever such a stimulus falls into the cell's receptive field (Lund et al., 1995). The optimum stimulus width for either cell type is, in the fovea, about two minutes of arc. The resolving power (acuity) of both cell types is the same.

About 10–20% of complex cells in the upper layers of the striate cortex show marked directional selectivity (Hubel, 1988). Directional Selectivity (DS) refers to the cell's response to a particular direction of movement. Cortical Directional Selectivity (CDS) contributes to motion perception and to the control of eye movements (Grzywacz & Norcia, 1995). CDS cells establish a motion pathway from V1 projecting to MT and V2 (which also projects to MT) and to MST. In contrast, there is no evidence that Retinal Directional Selectivity (RDS) contributes to motion perception. RDS contributes to oculomotor responses (Grzywacz et al., 1995). In vertebrates, it is involved in optokinetic nystagmus, a type of eye movement discussed in Chapter 4.

2.4.1 Motion-Sensitive Single-Cell Physiology

There are two somewhat counterintuitive implications of the visual system's motion-sensitive single-cell organization for perception. First, due to motion-sensitive cells, eye movements are never perfectly still but make constant tiny movements called *microsaccades* (Hubel, 1988). The counterintuitive fact regarding eye movements is that if an image were artificially stabilized on the retina, vision would fade away within about a second and the scene would become blank. Second, due to the response characteristics of single (cortical) cells, the cameralike "retinal buffer" representation of natural images is much more abstract than intuition suggests. An object in the visual field stimulates only a tiny fraction of the cells on whose receptive field it falls (Hubel, 1988). Perception of the object depends mostly on the response of (orientation-specific) cells to the object's borders. For example, the homogeneously shaded interior of an arbitrary form (e.g., a kidney bean) does not stimulate cells of the visual system. Awareness of the interior shade or hue depends on only cells sensitive to the borders of the object. In Hubel's (1988) words, "...our perception of the interior as black, white, gray, or green has nothing to do with cells whose fields are in the interior—hard as that may be to swallow.... What happens at the borders is the only information you need to know: the interior is boring."

2.5 Summary and Further Reading

This chapter presented a simplified view of the brain with emphasis on regions and structures of the brain responsible for attentional and visual processing, including those regions implicated in eye movement generation. Starting with the structure of the eye, the most salient observation is the structure of the retina which clearly shows the limited scope of the high resolution fovea. The division between foveo–peripheral vision is maintained along the visual pathways and can be clearly seen under microscope in the LGN. Of particular relevance to attention and eye movements is the physiological and functional duality of the magno- and parvocellular pathways and of their apparent mapping to their attentional "what" and "where" classification. Although this characterization of the M- and P-pathways is admittedly overly simplistic, it provides an intuitive functional distinction between foveal and peripheral vision.

An interesting visual example of foveo–peripheral processing is shown in Figure 2.7. To notice the curious difference between foveal and peripheral processing, foveate one corner of the image in Figure 2.7 and, without moving your eyes, shift your attention to the opposing corner of the image. Interestingly, you should perceive white dots at the line crossings in the foveal region, but black dots should appear at the line crossings in the periphery.

Examining regions in the brain along the visual pathways, one can obtain insight into how the brain processes visual information. The notion that attention may be driven by certain visual features (e.g., edges) is supported to an extent by the identification of neural regions which respond to these features. How certain features are perceived, particularly within and beyond the fovea, is the topic covered in the next chapter.

For an excellent review of physiological optics and visual perception in general, see Hendee and Wells (1997). For an introduction to neuroscience, see Hubel's (1988) very readable text. For a more recent description of the brain with an emphasis on color vision, see Zeki (1993). Apart from these texts on vision, several "handbooks" have also been assembled describing current knowledge of the brain. Arbib's (1995) handbook is one such example. It is an excellent source summarizing current knowledge of the brain, although it is somewhat difficult to read and to navigate through. [1] Another such well-organized but rather large text is Gazzaniga (2000).

---

[1] A new edition of Arbib's book has recently been announced.

Fig. 2.7. Foveo–peripheral illusion: scintillation effect produced by a variation of the standard Hermann grid illusion (attributed to L. Hermann (1870)), first discovered by Elke Lingelbach (at home). Adapted from Ninio and Stevens © 2000, Pion, London.

3

Visual Psychophysics

Given the underlying physiological substrate of the human visual system, measurable performance parameters often (but not always!) fall within ranges predicted by the limitations of the neurological substrate. Visual performance parameters, such as visual acuity, are often measured following established experimental paradigms, generally derived in the field of psychophysics (e.g., Receiver Operating Characteristics, or ROC paradigm, is one of the more popular experimental methods).

Unexpected observed visual performance is often a consequence of complex visual processes (e.g., visual illusions), or combinations of several factors. For example, the well-known Contrast Sensitivity Function, or CSF, describing the human visual system's response to stimuli of varying contrast and resolution, depends not only on the organization of the retinal mosaic, but also on the response characteristics of complex cellular combinations, e.g., receptive fields.

In this book, the primary concern is visual attention, and so the book primarily considers the distinction between foveo–peripheral vision. This subject, although complex, is discussed here in a fairly simplified manner, with the aim of elucidating only the most dramatic differences between what is perceived foveally and peripherally. In particular, visual (spatial) acuity is arguably the most studied distinction and is possibly the simplest parameter to alter in eye-based interaction systems (at least at this time). It is therefore the topic covered in greatest detail, in comparison to the other distinctions covered here briefly: temporal and chromatic foveo–peripheral differences.

3.1 Spatial Vision

Dimensions of retinal features are usually described in terms of projected scene dimensions in units of degrees visual angle, defined as

$$A = 2\arctan\frac{S}{2D},$$

where S is the size of the scene object and D is the distance to the object (see Figure 3.1). Common visual angles are given in Table 3.1.

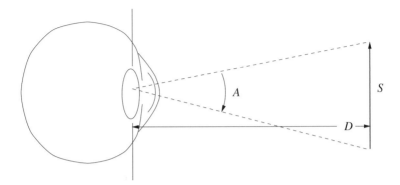

Fig. 3.1. Visual angle. Adapted from Haber and Hershenson (1973) © 1973. Reprinted with permission of Brooks/Cole, an imprint of the Wadsworth Group, a division of Thomson Learning.

Table 3.1. Common visual angles.

Object	Distance	Angle Subtended
Thumbnail	Arm's length	1.5–2°
Sun or moon	—	.5° or 30′ or arc
U.S. quarter coin	Arm's length	2°
U.S. quarter coin	85m	1′ (1 minute of arc)
U.S. quarter coin	5km	1″ (1 second of arc)

The innermost region is the fovea centralis (or foveola) which measures 400 μm in diameter and contains 25,000 cones. The fovea proper measures 1500 μm in diameter and holds 100,000 cones. The macula (or central retina) is 5000 μm in diameter, and contains 650,000 cones. One degree visual angle corresponds to approximately 300 μm distance on the human retina (De Valois & De Valois, 1988). The foveola, measuring 400 μm subtends 1.3° visual angle, and the fovea and macula subtend 5° and 16.7°, respectively (see Figure 3.2). Figure 3.3 shows the retinal distribution of rod and cone receptors. The fovea contains 147,000 cones/mm^2 and a slightly smaller number of rods. At about 10° the number of cones drops sharply to less than 20,000 cones/mm^2 and at 30° the number of rods in the periphery drops to about 100,000 rods/mm^2 (Haber & Hershenson, 1973).

The entire visual field roughly corresponds to a 23,400 square degree area defined by an ellipsoid with the horizontal major axis subtending 180° visual angle, and the

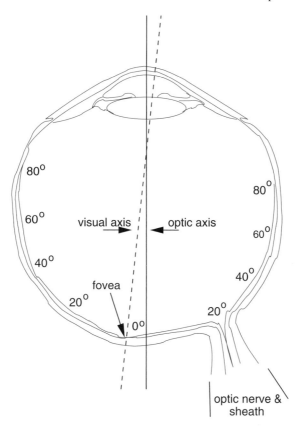

Fig. 3.2. Density distributions of rod and cone receptors across the retinal surface: visual angle. Adapted from Pirenne (1967); as cited in Haber and Hershenson (1973).

minor vertical axis subtending 130°. The diameter of the highest acuity circular region subtends 2°, the parafovea (zone of high acuity) extends to about 4° or 5°, and acuity drops off sharply beyond. At 5°, acuity is only 50% (Irwin, 1992). The so-called "useful" visual field extends to about 30°. The rest of the visual field has very poor resolvable power and is mostly used for perception of ambient motion. With increasing eccentricity the cones increase in size, whereas the rods do not (De Valois & De Valois, 1988). Cones, not rods, make the largest contribution to the information going to deeper brain centers, and provide most of the fine-grained spatial resolvability of the visual system.

The Modulation Transfer Function (MTF) theoretically describes the spatial resolvability of retinal photoreceptors by considering the cells as a finite array of sampling units. The 400 μm-diameter rod-free foveola contains 25,000 cones. Using the area of a circle, $25000 = \pi r^2$, approximately $2\sqrt{25000/\pi} = 178.41$ cones occupy a 400 μm linear cross-section of the foveola with an estimated average linear inter-cone

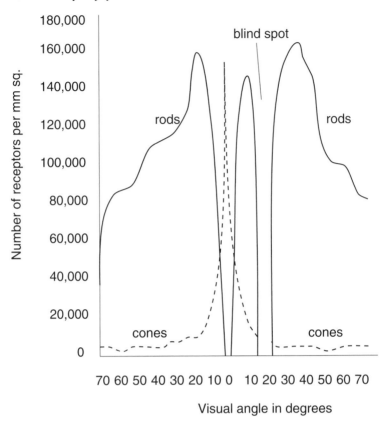

Fig. 3.3. Density distributions of rod and cone receptors across the retinal surface: rod/cone density. Adapted from Pirenne (1967); as cited in Haber and Hershenson (1973).

spacing of 2.24 μm. Cones in this region measure about 1 μm in diameter. Because one degree visual angle corresponds to approximately 300 μm distance on the human retina, roughly 133 cones are packed per degree visual angle in the foveola. By the sampling theorem, this suggests a resolvable spatial Nyquist frequency of 66 c/deg. Subjective resolution has in fact been measured at about 60 c/deg (De Valois & De Valois, 1988). In the fovea, a similar estimate based on the foveal diameter of 1500 μm and a 100,000 cone population, gives an approximate linear cone distribution of $2\sqrt{100000/\pi} = 356.82$ cones per 1500 μm. The average linear intercone spacing is then 71 cones/deg suggesting a maximum resolvable frequency of 35 cycles/deg, roughly half the resolvability within the foveola. This is somewhat of an underestimate because cone diameters increase twofold by the edge of the fovea suggesting a slightly milder acuity degradation. These one-dimensional approximations are not fully generalizable to the two-dimensional photoreceptor array although they provide insight into the theoretic resolution limits of the eye. Effective relative visual acuity measures are usually obtained through psychophysical experimentation.

At photopic light levels (day, or cone vision), foveal acuity is fairly constant within the central $2°$, and drops approximately linearly from there to the $5°$ foveal border. Beyond the $5°$, acuity drops sharply (approximately exponentially). At scotopic light levels (night, or rod-vision), acuity is poor at all eccentricities. Figure 3.4 shows the variation of visual acuity at various eccentricities and light intensity levels. Intensity is shown varying from 9.0 to 4.6 log micromicrolamberts, denoted by log mmL (9.0 log micromicrolamberts $= 10^9$ micromicrolamberts $= 1$ mL, see Davson (1980, p. 311)). The correspondence between foveal receptor spacing and optical limits generally holds in foveal regions of the retina, but not necessarily in the periphery. In contrast to the approximate 60 c/deg resolvability of foveal cones, the highest spatial frequencies resolvable by rods are on the order of 5 c/deg, suggesting poor resolvability in the relatively cone-free periphery. Although visual acuity correlates fairly well with cone distribution density, it is important to note that synaptic organization and later neural elements (e.g., ganglion cells concentrated in the central retina) are also contributing factors in determining visual acuity.

3.2 Temporal Vision

Human visual response to motion is characterized by two distinct facts: the *persistence of vision* and the *phi phenomenon* (Gregory, 1990). The former essentially describes the temporal sampling rate of the HVS, and the latter describes a threshold above which the HVS detects apparent movement. Both facts are exploited in television, cinema, and graphics to elicit perception of motion from successively displayed still images.

Persistence of vision describes the inability of the retina to sample rapidly changing intensities. A stimulus flashing at about 50–60 Hz (cycles per second) will appear steady (depending on contrast and luminance conditions and observers). This is known as the Critical Fusion Frequency (CFF).[1] A stylized representation of the CFF, based on measurements of response to temporal stimuli of varying contrast (i.e., a temporal contrast sensitivity function) is shown in Figure 3.5. Incidentally, the curve of the CFF resembles the shape of the curve of the Contrast Sensitivity Function (CSF) which describes retinal spatial frequency response. The CFF explains why flicker is not seen when viewing a sequence of (still) images at a high enough rate. The CFF illusion is maintained in cinema because frames are shown at 24 frames per second (fps, equivalent to Hz), but a three-bladed shutter raises the flicker rate to 72Hz (three for each picture). Television also achieves the CFF by displaying the signal at 60 fields per second. Television's analog to cinema's three-bladed shutter is the interlacing scheme: the typical television frame rate is about 30 frames per second (depending on the standard use, e.g., NTSC in North America, PAL in other

---

[1] Also sometimes referred to as the Critical Flicker Frequency.

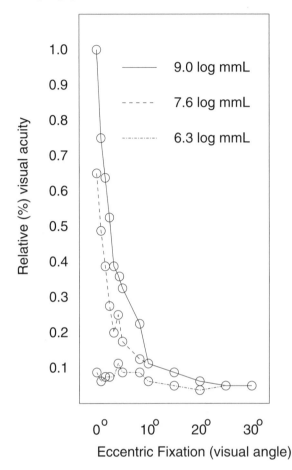

Fig. 3.4. Visual acuity at various eccentricities and light levels. Adapted from Davson (1980) with permission © 1980 Academic Press.

regions), but only the even or odd scanlines (fields) are shown per cycle. Although the CFF explains why flicker is effectively eliminated in motion picture (and computer) displays, it does not fully explain why motion is perceived.

The second fact that explains why movies, television, and graphics work is the phi phenomenon, or stroboscopic motion, or apparent motion. This fact explains the illusion of old-fashioned moving neon signs whose stationary lights are turned on in quick succession. This illusion can also be demonstrated with just two lights, provided the delay between successive light flashes is no less than about 62 Hz (Brinkmann, 1999). Inverting this value gives a rate of about 16 fps which is considered a bare minimum to facilitate the illusion of apparent motion.

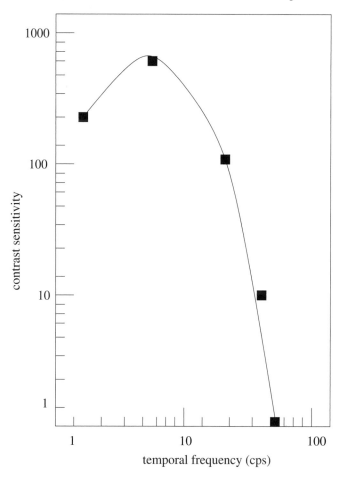

Fig. 3.5. Critical Fusion Frequency. Adapted from Bass (1995) © 1995 McGraw-Hill. Reproduced with permission of The McGraw-Hill Companies.

3.2.1 Perception of Motion in the Visual Periphery

In the context of visual attention and foveo–peripheral vision, the temporal response of the HVS is not homogeneous across the visual field. In terms of motion responsiveness, Koenderink et al. (1985) provide support that the foveal region is more receptive to slower motion than the periphery, although motion is perceived uniformly across the visual field. Sensitivity to target motion decreases monotonically with retinal eccentricity for slow and very slow motion (1 cycle/deg; Boff and Lincoln (1988)). That is, the velocity of a moving target appears slower in the periphery than in the fovea. Conversely, a higher rate of motion (e.g., frequency of rotation of grated disk) is needed in the periphery to match the apparent stimulus velocity in the fovea. At higher velocities, the effect is reversed.

Despite the decreased sensitivity in the periphery, movement is more salient there than in the central field of view (fovea). That is, the periphery is more sensitive to moving targets than to stationary ones. It is easier to peripherally detect a moving target than it is a stationary one. In essence, motion detection is the periphery's major task; it is a kind of early warning system for moving targets entering the visual field.

3.2.2 Sensitivity to Direction of Motion in the Visual Periphery

The periphery is approximately twice as sensitive to horizontal-axis movement as to vertical-axis movement (Boff & Lincoln, 1988). Directional motion sensitivity is show in Figure 3.6.

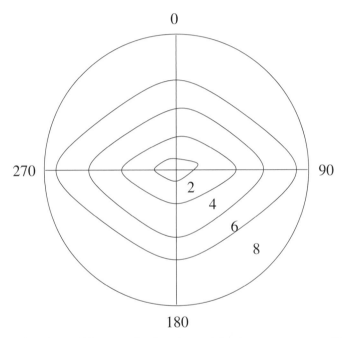

Fig. 3.6. Absolute threshold isograms for detecting peripheral rotary movement. Numbers are rates of pointer movement in revolutions per minute. Adapted from McColgin (1960) with permission © 1960 Optical Society of America.

3.3 Color Vision

Foveal color vision if facilitated by the three types of retinal cone photoreceptors. The three main spectral sensitivity curves for retinal cone photoreceptors peak at approximately 450 nm, 520 nm, and 555 nm wavelengths, for each of the blue, green, and red photoreceptors, respectively. A great deal is known about color vision in the fovea, however, relatively little is known about peripheral color vision. Of the seven million cones, most are packed tightly into the central 30° region of the fovea with scarcely any cones found beyond. This cone distribution suggests that peripheral color vision is quite poor in comparison to the color sensitivity of the central retinal region. Visual fields for monocular color vision are shown in Figure 3.7. Fields are shown for the right eye; fields for the left eye would be mirror images of those for the right eye. Blue and yellow fields are larger than the red and green fields; no chromatic visual fields have a definite border; instead, sensitivity drops off gradually and irregularly over a range of 15–30° visual angle (Boff & Lincoln, 1988).

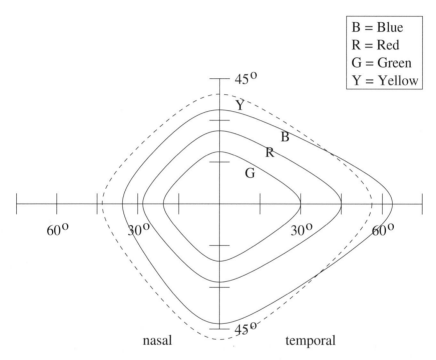

Fig. 3.7. Visual fields for monocular color vision (right eye). Adapted from Boff and Lincoln (1988) with permission ©1988 Wright-Patterson AFB.

Quantification of perceptual performance is not easily found in the literature. Compared to investigation of foveal color vision, only a few experiments have been performed to measure peripheral color sensitivity. Two studies, of particular relevance to peripheral location of color CRTs in an aircraft cockpit environment, investigated the chromatic discrimination of peripheral targets.

In the first study, Doyal (1991) concludes that peripheral color discrimination can approximate foveal discrimination when relatively small field sizes are presented (e.g., $2°$ at $10°$ eccentricity, and less than $4°$ at $25°$). Although this sounds encouraging, color discrimination was tested at limited peripheral eccentricities (within the central $30°$).

In the second, Ancman (1991) tested color discrimination at much greater eccentricities, up to about $80°$ visual angle. She found that subjects wrongly identified the color of a peripherally located $1.3°$ circle displayed on a CRT 5% of the time if it was blue, 63% of the time if red, and 62% of the time if green. Furthermore, blue could not be seen farther than $83.1°$ off the fovea (along the x-axis); red had to be closer than $76.3°$ and green nearer than $74.3°$ before subjects could identify the color.

There is much yet to be learned about peripheral color vision. Being able to verify a subject's direction of gaze during peripheral testing would be of significant benefit to these experiments. This type of psychophysical testing is but one of several research areas where eye tracking studies could play an important supporting role.

3.4 Implications for Attentional Design of Visual Displays

Both the structure and functionality of human visual system components place constraints on the design parameters of a visual communication system. In particular, the design of a gaze-contingent system must distinguish the characteristics of foveal and peripheral vision (see Section 20.2). A *visuotopic* representation model for imagery based on these observations is proposed:

1. *Spatial resolution* should remain high within the foveal region and smoothly degrade within the periphery, matching human visual acuity. High spatial frequency features in the periphery must be made visible "just in time" to anticipate gaze-contingent fixation changes.
2. *Temporal resolution* must be available in the periphery. Sudden onset events are potential attentional attractors. At low speeds, motion of peripheral targets should be increased to match apparent motion in the central field of view.
3. *Luminance* should be coded for high visibility in the peripheral areas because the periphery is sensitive to dim objects.
4. *Chrominance* should be coded for high exposure almost exclusively in the foveal region, with chromaticity decreasing sharply into the periphery. This requirement is a direct consequence of the high density of cones and parvocellular ganglion cells in the fovea.
5. *Contrast* sensitivity should be high in the periphery, corresponding to the sensitivity of the magnocellular ganglion cells found mainly outside the fovea.

Special consideration should be given to sudden onset, luminous, high-frequency objects (i.e., suddenly appearing bright edges).

A gaze-contingent visual system faces an implementational difficulty not yet addressed: matching the dynamics of human eye movement. Any system designed to incorporate an eye-slaved high resolution of interest, for example, must deal with the inherent delay imposed by the processing required to track and process real-time eye tracking data. To consider the temporal constraints that need to be met by such systems, the dynamics of human eye movements must be evaluated. This topic is considered in the following chapter.

3.5 Summary and Further Reading

Psychophysical information may be the most usable form of literature for the design of graphical displays, attentional in nature or otherwise. Introductory texts may include function plots of some aspect of vision (e.g., acuity) which may readily be used to guide the design of visual displays. However, one often needs to evaluate the experimental design used in psychophysical experiments to determine the generalizability of reported results. Furthermore, similar caution should be employed as in reading neurological literature: psychophysical results may often deal with a certain specific aspect of vision, which may or may not be readily applicable to display design. For example, visual acuity may suggest the use of relatively sized fonts on a Web page (larger font in the periphery), but acuity alone may not be sufficient to determine the required resolution in something like an attentional image or video display program. For the latter, one may need to piece together information concerning the visual contrast sensitivity function, temporal sensitivity, and so on. Furthermore, psychophysical studies may involve relatively simple stimuli (sine wave gratings), the results of which may or may not generalize to more complex stimuli such as imagery.

For a good introductory book on visual perception, see Hendee and Wells (1997). This text includes a good introductory chapter on the neurological basis of vision. Another good introductory book which also includes an interesting perspective on the perception of art is Solso (1999). For a somewhat terse but fairly complete psychophysical reference, see the USAF Engineering Data Compendium (Boff & Lincoln, 1988). This is an excellent "quick" guide to visual performance.

4

Taxonomy and Models of Eye Movements

Almost all normal primate eye movements used to reposition the fovea result as combinations of five basic types: saccadic, smooth pursuit, vergence, vestibular, and physiological nystagmus (miniature movements associated with fixations; Robinson (1968)). Vergence movements are used to focus the pair of eyes over a distant target (depth perception). Other movements such as adaptation and accommodation refer to nonpositional aspects of eye movements (i.e., pupil dilation, lens focusing). With respect to visual display design, positional eye movements are of primary importance.

4.1 The Extraocular Muscles and the Oculomotor Plant

In general, the eyes move within six degrees of freedom: three translations within the socket, and three rotations. There are six muscles responsible for movement of the eyeball: the *medial* and *lateral recti* (sideways movements), the *superior* and *inferior recti* (up/down movements), and the *superior* and *inferior obliques* (twist) (Davson, 1980). These are depicted in Figure 4.1. The neural system involved in generating eye movements is known as the oculomotor plant. The general plant structure and connections are shown in Figure 4.2 and described in Robinson (1968). Eye movement control signals emanate from several functionally distinct regions. Areas 17–19 and 22 are areas in the occipital cortex thought to be responsible for high-level visual functions such as recognition. The superior colliculus bears afferents emanating directly from the retina, particularly from peripheral regions conveyed through the magnocellular pathway. The semicircular canals react to head movements in three-dimensional space. All three areas (i.e., the occipital cortex, the superior colliculus, and the semicircular canals) convey efferents to the eye muscles through the mesencephalic and pontine reticular formations. Classification of observed eye movement signals relies in part on the known functional characteristics of these cortical regions.

Two pertinent observations regarding eye movements can be drawn from the oculomotor plant's organization:

Left (view from above): 1, superior rectus; 2, levator palbebrae superioris; 3, lateral rectus; 4, medial rectus; 5, superior oblique; 6, reflected tendon of the superior oblique; 7, annulus of Zinn. *Right (lateral view):* 8, inferior rectus; 9, inferior oblique.

Fig. 4.1. Extrinsic muscles of the eye. Adapted from Davson (1980) with permission © 1980 Academic Press.

1. The eye movement system is, to a large extent, a feedback circuit.
2. Signals controlling eye movement emanate from cortical regions that can be functionally categorized as voluntary (occipital cortex), involuntary (superior colliculus), and reflexive (semicircular canals).

The feedbacklike circuitry is utilized mainly in the types of eye movements requiring stabilization of the eye. Orbital equilibrium is necessitated for the steady retinal projection of an object, concomitant with the object's motion and movements of the head. Stability is maintained by a neuronal control system.

4.2 Saccades

Saccades are rapid eye movements used in repositioning the fovea to a new location in the visual environment. The term comes from an old French word meaning "flick of a sail" (Gregory, 1990). Saccadic movements are both voluntary and reflexive. The movements can be voluntarily executed or they can be invoked as a corrective optokinetic or vestibular measure (see below). Saccades range in duration from 10 ms to 100 ms, which is a sufficiently short duration to render the executor effectively blind during the transition (Shebilske & Fisher, 1983). There is some debate over the underlying neuronal system driving saccades. Saccades have been deemed ballistic and stereotyped. The term stereotyped refers to the observation that particular movement patterns can be evoked repeatedly. The term ballistic refers to the presumption that saccade destinations are preprogrammed. That is, once the saccadic movement to the next desired fixation location has been calculated (programming latencies of

CBT, corticobular tract; CER, cerebellum; ICTT, internal corticotectal tract; LG, lateral geniculate body; MLF, medial longitudinal fasciculus; MRF, mesencephalic and pontine reticular
formations; PT, pretectal nuclei; SA, stretch afferents from extraocular muscles; SC, superior
colliculi; SCC, semicircular canals; T, tegmental nuclei; VN, vestibular nuclei; II, optic nerve;
III, IV, and VI, the oculomotor, trochlear, and abducens nuclei and nerves; 17, 18, 19, 22, primary and association visual areas, occipital and parietal (Brodmann); 8, the frontal eye fields.

Fig. 4.2. Schematic of the major known elements of the oculomotor system. Adapted from
Robinson (1968) with permission © 1968 IEEE.

about 200 ms have been reported), saccades cannot be altered. One reason behind
this presumption is that, during saccade execution, there is insufficient time for visual feedback to guide the eye to its final position (Carpenter, 1977). One the other
hand, a saccadic feedback system is plausible if it is assumed that instead of visual
feedback, an internal copy of head, eye, and target position is used to guide the eyes
during a saccade (Laurutis & Robinson, 1986; Fuchs et al., 1985). Due to their fast

velocities, saccades may only appear to be ballistic (Zee et al., 1976).

Various models for saccadic programming have been proposed (Findlay, 1992). These models, with the exception of ones including "center-of-gravity" coding (see, e.g., He and Kowler (1989)), may inadequately predict unchangeable saccade paths. Instead, saccadic feedback systems based on an internal representation of target position may be more plausible because they tend to correctly predict the so-called double-step experimental paradigm. The double-step paradigm is an experiment where target position is changed during a saccade in midflight. Fuchs et al. (1985) proposed a refinement of Robinson's feedback model which is based on a signal provided by the superior colliculus and a local feedback loop. The local loop generates feedback in the form of motor error produced by subtracting eye position from a mental target-in-space position. Sparks and Mays (1990) cite compelling evidence that intermediate and deep layers of the SC contain neurons that are critical components of the neural circuitry initiating and controlling saccadic movements. These layers of the SC receive inputs from cortical regions involved in the analysis of sensory (visual, auditory, and somatosensory) signals used to guide saccades. The authors also rely on implications of Listing's and Donder's laws which specify an essentially null torsion component in eye movements, requiring virtually only two degrees of freedom for saccadic eye motions (Davson, 1980; Sparks & Mays, 1990). According to these laws, motions can be resolved into rotations about the horizontal x- and vertical y-axes.

Models of saccadic generation attempt to provide an explanation of the underlying mechanism responsible for generating the signals sent to the motor neurons. Although there is some debate as to the source of the saccadic program, the observed signal resembles a pulse/step function (Sparks & Mays, 1990). The pulse/step function refers to a dual velocity and position command to the extraocular muscles (Leigh & Zee, 1991). A possible simple representation of a saccadic step signal is a differentiation filter. Carpenter (1977) suggests such a possible filter arrangement for generating saccades coupled with an integrator. The integrating filter is in place to model the necessary conversion of velocity-coded information to position-coded signals (Leigh & Zee, 1991). A perfect neural integrator converts a pulse signal to a step function. An imperfect integrator (called leaky) will generate a signal resembling a decaying exponential function. The principle of this type of neural integration applies to all types of conjugate eye movements. Neural circuits connecting structures in the brain stem and the cerebellum exist to perform integration of coupled eye movements including saccades, smooth pursuits, and vestibular and optokinetic nystagmus (see below; Leigh and Zee (1991)).

A differentiation filter can be modeled by a linear filter as shown in Figure 4.3. In the time domain, the linear filter is modeled by the following equation

$$x_t = g_0 s_t + g_1 s_{t-1} + \cdots$$

$$= \sum_{k=0}^{\infty} g_k s_{t-k},$$

where s_t is the input (pulse), x_t is the output (step), and g_k are the filter coefficients. To ensure differentiation, the filter coefficients typically must satisfy properties that approximate mathematical differentiation. An example of such a filter is the Haar filter with coefficients $\{1, -1\}$. Under the z-transform the transfer function $X(z)/S(z)$ of this linear filter is

$$
\begin{aligned}
x_t &= g_0 s_t + g_1 s_{t-1} \\
x_t &= (1)s_t + (-1)s_{t-1} \\
x_t &= (1)s_t + (-1)z s_t \\
x_t &= (1-z)s_t \\
X(z) &= (1-z)S(z) \\
\frac{X(z)}{S(z)} &= 1 - z.
\end{aligned}
$$

The Haar filter is a length-2 filter which approximates the first derivate between successive pairs of inputs.

linear moving average system

input   filter   output

s_t g x_t

Fig. 4.3. Diagram of a simple linear filter modeling saccadic movements.

4.3 Smooth Pursuits

Pursuit movements are involved when visually tracking a moving target. Depending on the range of target motion, the eyes are capable of matching the velocity of the moving target. Pursuit movements provide an example of a control system with built-in negative feedback (Carpenter, 1977). A simple closed-loop feedback loop used to model pursuit movements is shown in Figure 4.4, where s_t is the target position, x_t is the (desired) eye position, and h is the (linear, time-invariant) filter, or gain of the system (Carpenter, 1977; Leigh & Zee, 1991). Tracing the loop from the feedback start point gives the following equation in the time domain

$$h(s_t - x_t) = x_{t+1}.$$

Under the z-transform the transfer function $X(z)/S(z)$ of this linear system is

$$H(z)(S(z) - X(z)) = X(z)$$
$$H(z)S(z) = X(z)(1 + H(z))$$
$$\frac{H(z)}{1 + H(z)} = \frac{X(z)}{S(z)}.$$

In the closed-loop feedback model, signals from visual receptors constitute the error signal indicating needed compensation to match the target's retinal image motion.

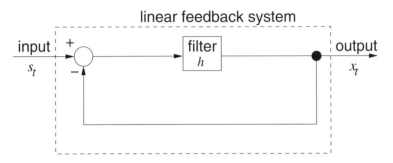

Fig. 4.4. Diagram of a simple linear feedback model of smooth pursuit movements.

4.4 Fixations

Fixations are eye movements that stabilize the retina over a stationary object of interest. It seems intuitive that fixations should be generated by the same neuronal circuit controlling smooth pursuits with fixations being a special case of a target moving at zero velocity. This is probably incorrect (Leigh & Zee, 1991, pp.139–140). Fixations, instead, are characterized by the miniature eye movements: tremor, drift, and microsaccades. This is a somewhat counterintuitive consequence of the visual system's motion-sensitive single-cell organization. Recall that microsaccades are made due to the motion sensitivity of the visual system's single-cell physiology. Microsaccades are eye movement signals that are more or less spatially random varying over 1 to 2 minutes of arc in amplitude. The counterintuitive fact regarding fixations is that if an image is artificially stabilized on the retina, vision fades away within about a second and the scene becomes blank.

Miniature eye movements that effectively characterize fixations may be considered noise present in the control system (possibly distinct from the smooth pursuit circuit) attempting to hold gaze steady. This noise appears as a random fluctuation about the area of fixation, typically no larger than 5° visual angle (Carpenter, 1977, p.105). Although the classification of miniature movements as noise may be an oversimplification of the underlying natural process, it allows the signal to be modeled by a feedback system similar to the one shown in Figure 4.4. The additive noise in Figure 4.4 is represented by $e_t = s_t - x_t$, where the (desired) eye position x_t is subtracted from the steady fixation position s_t at the summing junction. In this model, the error signal stimulates the fixation system in a manner similar to the smooth pursuit system, except that here e_t is an error-position signal instead of an error-velocity signal (see Leigh and Zee (1991, p.150)). The feedback system modeling fixations, using the noisy "data reduction" method, is in fact simpler than the pursuit model because it implicitly assumes a stationary stochastic process (Carpenter, 1977, p.107). Stationarity in the statistical sense refers to a process with constant mean. Other relevant statistical measures of fixations include their duration range of 150 ms to 600 ms, and the observation that 90% of viewing time is devoted to fixations (Irwin, 1992).

4.5 Nystagmus

Nystagmus eye movements are conjugate eye movements characterized by a saw-toothlike time course (time series signal) pattern. Optokinetic nystagmus is a smooth pursuit movement interspersed with saccades invoked to compensate for the retinal movement of the target. The smooth pursuit component of optokinetic nystagmus appears in the slow phase of the signal (Robinson, 1968). Vestibular nystagmus is a similar type of eye movement compensating for the movement of the head. The time course of vestibular nystagmus is virtually indistinguishable from its optokinetic counterpart (Carpenter, 1977).

4.6 Implications for Eye Movement Analysis

From the above discussion, two significant observations relevant to eye movement analysis can be made. First, based on the functionality of eye movements, only three types of movements need be modeled to gain insight into the overt localization of visual attention. These types of eye movements are fixations, smooth pursuits, and saccades. Second, based on signal characteristics and plausible underlying neural circuitry, all three types of eye movements may be approximated by a Linear, Time-Invariant (LTI) system (i.e., a linear filter; for examples of linear filters applicable to saccade detection, see Chapter 12).

The primary requirement of eye movement analysis, in the context of gaze-contingent system design, is the identification of fixations, saccades, and smooth pursuits. It is assumed that these movements provide evidence of voluntary, overt visual attention. This assumption does not preclude the plausible involuntary utility of these movements, or conversely, the covert nonuse of these eye movements (e.g., as in the case of parafoveal attention). Fixations naturally correspond to the desire to maintain one's gaze on an object of interest. Similarly, pursuits are used in the same manner for objects in smooth motion. Saccades are considered manifestations of the desire to voluntarily change the focus of attention.

4.7 Summary and Further Reading

This chapter presented a taxonomy of eye movements and included linear models of eye movement signals suitable for eye movement analysis (see also Chapter 12).

With the exception of Carpenter's widely referenced text (Carpenter, 1977), there appears to be no single suitable introductory text discussing eye movements exclusively. Instead, there are various texts on perception, cognition, and neuroscience which often include a chapter or section on the topic. There are also various collections of technical papers on eye movements, usually assembled from proceedings of focused symposia or conferences. A series of such books was produced by John Senders et al. in the 1970s and 1980s (see for, e.g., Monty and Senders (1976) and Fisher et al. (1981)). This conference series has recently been revived in the form of the (currently biennial) Eye Tracking Research & Applications (ETRA) conference.

A large amount of work has been performed on studying eye movements in the context of reading. For a good introduction to this literature, see Rayner (1992).

Part II

Eye Tracking Systems

5

Eye Tracking Techniques

The measurement device most often used for measuring eye movements is commonly known as an eye tracker. In general, there are two types of eye movement monitoring techniques: those that measure the position of the eye relative to the head, and those that measure the orientation of the eye in space, or the "point of regard" (Young & Sheena, 1975). The latter measurement is typically used when the concern is the identification of elements in a visual scene, e.g., in (graphical) interactive applications. Possibly the most widely applied apparatus for measurement of the point of regard is the video-based corneal reflection eye tracker. In this chapter, most of the popular eye movement measurement techniques are briefly discussed first before covering video-based trackers in greater detail.

There are four broad categories of eye movement measurement methodologies involving the use or measurement of: Electro-OculoGraphy (EOG), scleral contact lens/search coil, Photo-OculoGraphy (POG) or Video-OculoGraphy (VOG), and video-based combined pupil and corneal reflection.

Electro-oculography relies on (d.c. signal) recordings of the electric potential differences of the skin surrounding the ocular cavity. During the mid-1970s, this technique was the most widely applied eye movement method (Young & Sheena, 1975). Today, possibly the most widely applied eye movement technique, primarily used for point of regard measurements, is the method based on corneal reflection.

The first method for objective eye measurements using corneal reflection was reported in 1901 (Robinson, 1968). To improve accuracy, techniques using a contact lens were developed in the 1950s. Devices attached to the contact lens ranged from small mirrors to coils of wire. Measurement devices relying on physical contact with the eyeball generally provide very sensitive measurements. The obvious drawback of these devices is their invasive requirement of wearing the contact lens. So-called non-invasive (sometimes called remote) eye trackers typically rely on the measurement of visible features of the eye, e.g., the pupil, iris–sclera boundary, or a corneal reflection of a closely positioned, directed light source. These techniques often involve either manual or automatic (computer-based) analysis of video recordings of the movements of the eyes, either off-line or in real-time. The availability of fast image processing hardware has facilitated the development of real-time video-based point of regard turnkey systems.

5.1 Electro-OculoGraphy (EOG)

Electro-oculography, the most widely applied eye movement recording method some 40 years ago (and still used today), relies on measurement of the skin's electric potential differences, of electrodes placed around the eye. A picture of a subject wearing the EOG apparatus is shown in Figure 5.1. The recorded potentials are in the range 15–200 μV, with nominal sensitivities of order of 20 μV/deg of eye movement. This technique measures eye movements relative to head position, and so is not generally suitable for point of regard measurements unless head position is also measured (e.g., using a head tracker).

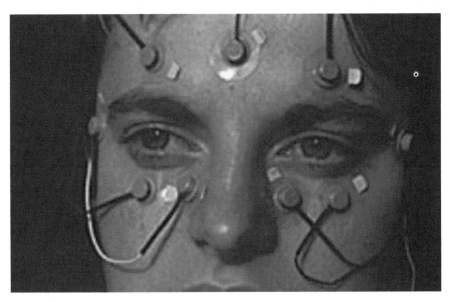

Fig. 5.1. Example of electro-oculography (EOG) eye movement measurement. Courtesy of MetroVision, Pérenchies, France <http://www.metrovision.fr>. Reproduced with permission.

5.2 Scleral Contact Lens/Search Coil

One of the most precise eye movement measurement methods involves attaching a mechanical or optical reference object mounted on a contact lens which is then worn directly on the eye. Such early recordings (ca. 1898; Young and Sheena (1975)) used a plaster of paris ring attached directly to the cornea and through mechanical linkages to recording pens. This technique evolved to the use of a modern contact lens to which a mounting stalk is attached. The contact lens is necessarily large, extending over the cornea and sclera (the lens is subject to slippage if the lens only covers the cornea). Various mechanical or optical devices have been placed on the stalk attached to the lens: reflecting phosphors, line diagrams, and wire coils have been the most popular implements in magneto-optical configurations. The principle method employs a wire coil, which is then measured moving through an electromagnetic field.[1] A picture of the search coil embedded in a scleral contact lens and the electromagnetic field frame are shown in Figure 5.2. The manner of insertion of the contact lens is shown in Figure 5.3. Although the scleral search coil is the most precise eye movement measurement method (accurate to about 5–10 arc-seconds over a limited range of about 5°; Young and Sheena (1975)), it is also the most intrusive method. Insertion of the lens requires care and practice. Wearing of the lens causes discomfort. This method also measures eye position relative to the head, and is not generally suitable for point of regard measurement.

Fig. 5.2. Example of search coil embedded in contact lens and electromagnetic field frames for search coil eye movement measurement. Courtesy of Skalar Medical, Delft, The Netherlands <http://www.skalar.nl>. Reproduced with permission.

5.3 Photo-OculoGraphy (POG) or Video-OculoGraphy (VOG)

This category groups together a wide variety of eye movement recording techniques involving the measurement of distinguishable features of the eyes under rotation/-translation, e.g., the apparent shape of the pupil, the position of the limbus (the iris-sclera boundary), and corneal reflections of a closely situated directed light source

---

[1] This is similar in principle to magnetic position/orientation trackers often employed in virtual reality applications; e.g., Ascension's Flock Of Birds (FOB) uses this type of method for tracking the position/orientation of the head. See Chapter 7.

Fig. 5.3. Example of scleral suction ring insertion for search coil eye movement measurement. Courtesy of Skalar Medical, Delft, The Netherlands <http://www.skalar.nl>. Reproduced with permission.

(often infra-red). Although different in approach, these techniques are grouped here because they often do not provide point of regard measurement. Examples of apparatus and recorded images of the eye used in photo- or video-oculography and/or limbus tracking are shown in Figure 5.4. Measurement of ocular features provided by these measurement techniques may or may not be made automatically, and may involve visual inspection of recorded eye movements (typically recorded on videotape). Visual assessment performed manually (e.g., stepping through a videotape frame-by-frame), can be extremely tedious and prone to error, and limited to the temporal sampling rate of the video device.

Automatic limbus tracking often involves the use of photodiodes mounted on spectacle frames (see Figure 5.4b and c), and almost always involves the use of invisible (usually infra-red) illumination (see Figure 5.4(d)). Several of these methods require the head to be fixed, e.g., either by using a head/chin rest or a bite bar (Young & Sheena, 1975).

5.4 Video-Based Combined Pupil/Corneal Reflection

Although the above techniques are in general suitable for eye movement measurements, they do not often provide point of regard measurement. To provide this measurement, either the head must be fixed so that the eye's position relative to the head and point of regard coincide, or multiple ocular features must be measured in order to disambiguate head movement from eye rotation. Two such features are the corneal reflection (of a light source, usually infra-red) and the pupil center (see Figure 5.4d).

Video-based trackers utilize relatively inexpensive cameras and image processing hardware to compute the point of regard in real-time. The apparatus may be table-mounted, as shown in Figure 5.5 or worn on the head, as shown in Figure 5.6. The optics of both table-mounted or head-mounted systems are essentially identical, with the exception of size. These devices, which are becoming increasingly available, are most suitable for use in interactive systems.

The corneal reflection of the light source (typically infra-red) is measured relative to the location of the pupil center. Corneal reflections are known as the Purkinje

(a) Example of apparent pupil size. Courtesy of MetroVision, Pérenchies, France <http://www.metrovision.fr>. Reproduced with permission.

(b) Example of infra-red limbus tracker apparatus. Courtesy of Applied Science Laboratories (ASL), Bedford, MA <http://www.a-s-l.com>. Reproduced with permission.

(c) Another example of infra-red limbus tracker apparatus, as worn by subject. Courtesy of Microguide, Downers Grove, IL <http://www.eyemove.com>. Reproduced with permission.

(d) Example of "bright pupil" (and corneal reflection) illuminated by infra-red light. Courtesy of LC Technologies, Fairfax, VA <http://www.eyegaze.com>. Reproduced with permission.

Fig. 5.4. Examples of pupil, limbus, and corneal infra-red (IR) reflection eye movement measurements.

(a) Operator.   (b) Subject.

Fig. 5.5. Example of table-mounted video-based eye tracker.

Fig. 5.6. Example of head-mounted video-based eye tracker. Courtesy of IOTA AB, EyeTrace Systems, Sundsvall Business & Tech. Center, Sundsvall, Sweden <http://www.iota.se>. Reproduced with permission.

reflections, or Purkinje images (Crane, 1994). Due to the construction of the eye, four Purkinje reflections are formed, as shown in Figure 5.7. Video-based eye trackers typically locate the first Purkinje image. With appropriate calibration procedures, these eye trackers are capable of measuring a viewer's Point Of Regard (POR) on a suitably positioned (perpendicularly planar) surface on which calibration points are displayed.

Two points of reference on the eye are needed to separate eye movements from head movements. The positional difference between the pupil center and corneal reflection changes with pure eye rotation, but remains relatively constant with minor head movements. Approximate relative positions of pupil and first Purkinje reflections are graphically shown in Figure 5.8, as the left eye rotates to fixate nine correspondingly placed calibration points. The Purkinje reflection is shown as a small white circle in close proximity to the pupil, represented by a black circle. Because the infra-red light source is usually placed at some fixed position relative to the eye, the

PR, Purkinje reflections: 1, reflection from front surface of the cornea; 2, reflection from rear surface of the cornea; 3, reflection from front surface of the lens; 4, reflection from rear surface of the lens–almost the same size and formed in the same plane as the first Purkinje image, but due to change in index of refraction at rear of lens, intensity is less than 1% of that of the first Purkinje image; IL, incoming light; A, aqueous humor; C, cornea; S, sclera; V, vitreous humor; I, iris; L, lens; CR, center of rotation; EA, eye axis; $a \approx 6$ mm; $b \approx 12.5$ mm; $c \approx 13.5$ mm; $d \approx 24$ mm; $r \approx 7.8$ mm (Crane, 1994).

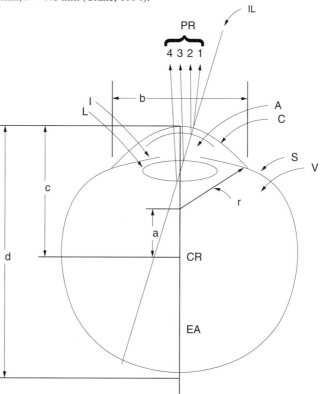

Fig. 5.7. Purkinje images. Reprinted in adapted form from Crane (1994) courtesy of Marcel Dekker, Inc. © 1994.

Purkinje image is relatively stable whereas the eyeball, and hence thepupil, rotates in its orbit. So-called generation-V eye trackers also measure the fourth Purkinje image (Crane & Steele, 1985). By measuring the first and fourth Purkinje reflections, these dual-Purkinje image (DPI) eye trackers separate translational and rotational eye movements. Both reflections move together through exactly the same distance upon eye translation but the images move through different distances (thus changing their separation) upon eye rotation. This device is shown in Figure 5.9. Unfortunately, although the DPI eye tracker is quite precise, head stabilization may be required.

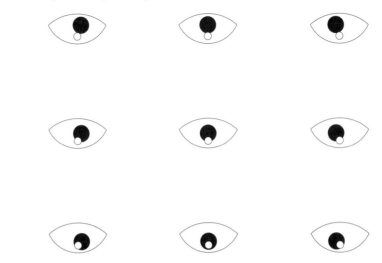

Fig. 5.8. Relative positions of pupil and first Purkinje images as seen by the eye tracker's camera.

Fig. 5.9. Dual-Purkinje eye tracker. Courtesy of Fourward Optical Technologies, Buena Vista, VA <http://www.fourward.com>. Reproduced with permission.

5.5 Classifying Eye Trackers in "Mocap" Terminology

For readers familiar with motion capture ("mocap") techniques used in the special effects film industry, it is worthwhile to compare the various eye tracking methodologies with traditional mocap devices. Similarities between the two applications are intuitive and this is not surprising because the objective of both is recording the motion of objects in space. In eye tracking, the object measured is the eye, whereas in mocap, it is (usually) the joints of the body. Eye trackers can be grouped using the same classification employed to describe motion capture devices.

Electro-OculoGraphy (EOG) is essentially an electromechanical device. In mocap applications, sensors may be placed on the skin or joints. In eye tracking, sensors are placed on the skin around the eye cavity. Eye trackers using a contact lens are effectively electromagnetic trackers. The metallic stalk that is fixed to the contact lens is similar to the orthogonal coils of wire found in electromagnetic sensors used to obtain the position and orientation of limbs and head in virtual reality. Photo-oculography and video-oculography eye trackers are similar to the widely used optical motion capture devices in special effects film, video, and game production. In both cases a camera is used to record raw motion, which is then processed by (usually) digital means to calculate the motion path of the object being tracked. Finally, video-based corneal reflection eye trackers are similar to optical motion capture devices that use reflective markers (worn by the actors). In both cases, an infra-red light source is usually used, for the reason that it is invisible to the human eye, and hence nondistracting.

For a good introduction to motion capture for computer animation, as used in special effects film, video, and game production, see Menache (2000). Menache's book does a good job of describing motion capture techniques, although it is primarily aimed at practitioners in the field of special effects and animation production. Still, the description of mocap techniques, even at a comparatively large scale (i.e., capturing human motion vs. motion of the eye), provides a good classification scheme for eye tracking techniques.

5.6 Summary and Further Reading

For a short review of early eye tracking methods, see Robinson (1968, II). For another relatively more recent survey of eye tracking techniques, see Young and Sheena (1975). Although Young and Sheena's survey article is somewhat dated by today's standards, it is an excellent introductory article on eye movement recording techniques, and is still widely referenced. An up-to-date survey of eye tracking devices does not appear to be available, although the number of (video-based) eye tracking companies seems to be growing. Instead, one of the best comprehensive lists of eye tracking manufacturers is available on the Internet, the Eye Movement Equipment Database (EMED; see <http://ibs.derby.ac.uk/emed/>), initially setup by David Wooding of the Institute of Behavioural Sciences, University of Derby, United Kingdom.

6

Head-Mounted System Hardware Installation

Several types of eye trackers are available, ranging from scleral coils, EOG, to video-based corneal reflection eye trackers. Although each has its benefits and drawbacks (e.g., accuracy vs. sampling rate), for graphical or interactive applications the video-based corneal reflection tracker is arguably the most practical device. These devices work by capturing video images of the eye (illuminated by an infra-red light source), processing the video frames (at video frame rates), and outputting the eye's x- and y-coordinates relative to the screen being viewed. The x- and y-coordinates are typically either stored by the eye tracker itself, or can be sent to the graphics host via serial cable. The advantage of the video-based eye tracker over other devices is that it is relatively noninvasive, fairly accurate (to about $1°$ visual angle over a $30°$ viewing range), and, for the most part, not difficult to integrate with the graphics system. The video-based tracker's chief limitation is its sampling frequency, typically limited by the video frame rate, 60 Hz. Hence, one can usually expect to receive eye movement samples at least every 16 ms (typically a greater latency should be expected because the eye tracker needs time to process each video frame, and the graphics host needs time to update its display).

6.1 Integration Issues and Requirements

Integration of the eye tracker into a graphics system depends chiefly on proper delivery of the graphics video stream to the eye tracker and the subsequent reception of the eye tracker's calculated 2D eye coordinates. In the following description of system setup, a complete graphics system is described featuring two eye trackers: one a table-mounted, monocular eye tracker set underneath a standard television display, and the other a binocular eye tracker fitted inside a Head-Mounted Display (HMD). Both displays are powered by the same graphics host. Such a laboratory has been set up within Clemson University's Virtual Reality Laboratory, and is shown in Figure 6.1. In Figure 6.1b, the portion of the lab is shown where, on the right, the TV display is installed with the monocular table-mounted eye tracker positioned below. In Figure 6.1a, the HMD helmet is resting atop a small table in front of three d.c. electromagnetic tracking units. The eye tracker unit (a PC) is situated between the d.c. tracking units and the dual-head graphics display monitors to the right in the image. The PC's display is a small flat panel display just left of the dual graphics display monitors. Both HMD and TV (and graphics) displays are driven by the graphics host, which is out of view of the image (on the floor beneath the desk in front of the visible chair). The four small TV monitors atop the eye tracking PC in Figure 6.1a display the left and right scene images (what the user sees) and the left and right eye images (what the eye tracker sees).

The following system integration description is based on the particular hardware devices installed at Clemson's Virtual Reality Eye Tracking (VRET) laboratory, described here for reference. The primary rendering engine is a Dell Dual 1.8 GHz Pentium 4 (Xeon) PC (1 GB RAM) running Fedora Core 4 and equipped with an nVidia GeForce4 FX5950U graphics card. The eye tracker is from ISCAN and the system includes both binocular cameras mounted within a Virtual Research V8 (high resolution) HMD as well as a monocular camera mounted on a remote pan-tilt unit. Both sets of optics function identically, capturing video images of the eye and sending the video signal to the eye tracking PC for processing. The pan-tilt unit coupled with the remote table-mounted camera/light assembly is used to noninvasively track the eye in real-time as the subject's head moves. This allows limited head movement, but a chin rest is usually used to restrict the position of the subject's head during experimentation to improve accuracy. The V8 HMD offers 640 × 480 resolution per eye with separate left and right eye feeds. HMD position and orientation tracking is provided by an Ascension 6 Degree-Of-Freedom (6 DOF) Flock Of Birds (FOB), a d.c. electromagnetic system with a 10 ms latency per sensor. A 6 DOF tracked, handheld mouse provides the user with directional motion control. The HMD is equipped with headphones for audio localization.

Although the following integration and installation guidelines are based on the equipment available at the Clemson VRET lab, the instructions should apply to practically any video-based corneal reflection eye tracker. Of primary importance to proper system integration are the following,

1. Knowledge of the video format the eye tracker requires as input (e.g., NTSC or VGA)
2. Knowledge of the data format the eye tracker generates as its output

The first point is crucial to providing the proper image to the user as well as to the eye tracker. The eye tracker requires input of the scene signal so that it can overlay the calculated point of regard for display on the scene monitor which is viewed by the operator (experimenter). The second requirement is needed for proper interpretation of the point of regard by the host system. This is usually provided by the eye tracker manufacturer. The host system will need to be furnished with a device driver to read and interpret this information.

Secondary requirements for proper integration are the following.

(a) HMD-mounted binocular eye tracker (panorama photostitch image).

(b) Table-mounted monocular eye tracker.

Fig. 6.1. Virtual Reality Eye Tracking (VRET) lab at Clemson University.

1. The capability of the eye tracker to provide fine-grained cursor control over its calibration stimulus (a crosshair or other symbol)
2. The capability of the eye tracker to transmit its operating mode to the host along with the eye x- and y-point of regard information

These two points are both related to proper alignment of the graphics display with the eye tracking scene image and calibration point. Essentially, both eye tracker capabilities are required to properly map between the graphics and eye tracking viewport coordinates. This alignment is carried out in two stages. First, the operator can use the eye tracker's own calibration stimulus to read out the extents of the eye tracker's displays. The eye tracker resolution may be specified over a 512×512 range, however, in practice it may be difficult to generate a graphics window that will exactly match the dimensions of the video display. Differences on the order of one or two pixels will ruin proper alignment. Therefore, it is good practice to first display a blank graphics window, then use the eye tracker cursor to measure the extents of the window in the eye tracker's reference frame. Because this measurement should be as precise as possible, fine cursor control is needed. Second, the eye tracker operates in three primary modes: reset (inactive), calibration, and run. The graphics program must synchronize with these modes, so that a proper display can be generated in each mode:

- *Reset*: display nothing (black screen) or a single calibration point.
- *Calibration*: display a simple small stimulus for calibration (e.g., a small dot or circle) at the position of the eye tracker's calibration stimulus.
- *Run*: display the stimulus required over which eye movements are to be recorded.

It is of the utmost importance that proper alignment is achieved between the eye tracker's calibration stimulus points and those of the graphics' display. Without this alignment, the data returned by the eye tracker will be meaningless.

Proper alignment of the graphics and eye tracking reference frames is achieved through a simple linear mapping between eye tracking and graphics window coordinates. This is discussed in Chapter 7, following notes on eye tracking system installation and wiring.

6.2 System Installation

The two primary installation considerations are wiring of the video and serial line cables between the graphics host and eye tracking systems. The connection of the serial cable is comparatively simple. Generation of the software driver for data interpretation is also straightforward, and is usually facilitated by the vendor's description of the data format and transmission rate. In contrast, what initially may pose the most problems is the connection of the video signal. It is imperative that the graphics host can generate a video signal in the format expected by both the graphics display (e.g., television set or HMD unit) and the eye tracker.

In the simplest case, if the host computer is capable of generating a video signal that is suitable for both the stimulus display and eye tracker, all that is then needed is a video splitter which will feed into both the stimulus display and eye tracker. For example, assume that the stimulus display is driven by an NTSC signal (e.g., a television), and the host computer is capable of generating a display signal in this format. (This was possible at the Clemson VRET lab because the SGI host can send a copy of whatever is on the graphics display via proper use of the ircombine command.) If the eye tracker can also be driven by an NTSC signal, then the installation is straightforward. If, however, the stimulus display is driven by VGA video (e.g., an HMD), but the eye tracker is driven by NTSC video, then the matter is somewhat more complicated. Consider the wiring diagram given in Figure 6.2. This schematic shows the dual display components, the HMD and TV, used for binocular eye tracking in a virtual reality environment, and monocular eye tracking over a 2D image or video display. The diagram features the necessary wiring of both left and right video channels from the host to the HMD and eye tracker, and a copy of the left video channel sent to the TV through the host's NTSC video output.

Fig. 6.2. Video signal wiring of the VRET lab at Clemson University.

The HMD is driven by a horizontal-synch (h-sync) VGA signal. A switchbox is used (seen in Figure 6.1a just above the eye tracker keyboard) to switch the VGA display between the dual-head graphics monitors and the HMD. The HMD video control box diverts a copy of the left video channel through an active pass-through splitter back through the switchbox to the left graphics display monitor. The switchbox effectively "steals" the signal meant for the graphics displays and sends it to the HMD. The left switch on the switchbox has two settings: monitor or HMD. The right switch on the switchbox has three settings: monitor, HMD, or both. If the right switch is set to monitor, no signal is sent to the HMD, effectively providing a biocular display in the HMD (instead of a binocular, or stereoscopic display). If the right switch is set to HMD, the graphics display blanks out because the HMD does not provide an active pass-through of the right video channel. If the right switch is set to both, the right video channel is simply split between the HMD and the monitor, resulting in a binocular display in both the HMD and on the monitors. This last setting provides no amplification of the signal and hence both the right LCD in the HMD and the right graphics monitor displays appear dim. This is mostly used for testing purposes.

The entire video circuit between the graphics host, the switchbox, and the HMD is VGA video. The eye tracker, however, operates on NTSC. This is the reason for the two VGA/NTSC converters that are inserted into the video path. These converters output an NTSC signal to the eye tracker and also provide active pass-throughs for the VGA signal so that, when in operation, the VGA signal appears undisrupted. The eye tracker then processes the video signals of the scene and outputs the signal to the scene monitors, with its own overlaid signal containing the calculated point of regard (represented by a crosshair cursor). These two small displays show the operator what is in the user's field of view as well as what she or he is looking at.

The eye images, in general, do not pose a complication since this video signal is exclusively processed by the eye tracker. In the case of the eye tracker at Clemson's VRET lab, both the binocular HMD eye cameras and the table-mounted monocular camera are NTSC and the signals feed directly into the eye tracker.

6.3 Lessons Learned from the Installation at Clemson

The eye tracker at Clemson contains hardware to process dual left and right eye and scene images. It can be switched to operate in monocular mode, for which it requires just the left eye and scene images. In this case, a simple video switch is used to switch the signal between the eye image generated by the left camera in the HMD and the camera on the table-mounted unit.

The first installation at Clemson used display monitors from Silicon Graphics, Inc. (SGI) that were driven by either VGA video, or the default R, G, B video delivered by 13W3 video cables. To drive the HMD, VGA video was required, connected by HD15 cables. To connect the video devices properly, special 13W3-HD15 cables were needed. Although this seems trivial, custom-built cables were required. These cables were not cheap, and took a day or two to build and deliver. If timing and finances are a consideration, planning of the system down to the proper cabling is a must! (Today this is less of a concern although properly matching cabling is still an issue, this time not so much concerning the video signal as concerning the Keyboard Video Monitor or KVM switch; see Chapter 9.)

A problem that was particularly difficult to troubleshoot during the Clemson installation was the diagnosis of the format of the VGA signal emitted by the SGI host computer. Initially, before the eye tracker was installed, the HMD was tested for proper display. The output of the switchbox was used directly to drive the HMD. Everything functioned properly. However, after inserting the HMD into the video circuit, the eye tracker would not work. It was later found that the problem lay in the VGA/NTSC converters: these converters expect the more common VGA signal which uses a timing signal synchronized to the horizontal video field (h-sync signal; the horizontal and vertical sync signals are found on pins 13 and 14 of the VGA HD15 cable). The SGI host computer by default emits a sync-on-green VGA signal leaving pins 13 and 14 devoid of a timing signal. The VR HMD contains circuitry that will read either h-sync or sync-on-green VGA video and so functions quietly given either signal. The fault, as it turned out, was in an improper initial wiring of the switchbox. The switchbox was initially missing connections for pins 13 and 14 because these are not needed by a sync-on-green VGA signal. Once this was realized, the entire cable installation had to be disassembled and the switchbox had to be rewired. With the switchbox rewired, the custom 13W3 cables had to be inspected to confirm that these components carried signals over pins 13 and 14. Finally, a new display configuration had to be created (using SGI's **ircombine** command) to drive the entire circuit with the horizontal sync signal instead of the default sync-on-green. The moral here is: be sure of the video format required by all components, down to the specific signals transmitted on individual cable pins!

6.4 Summary and Further Reading

This chapter presented key points from an eye tracker's installation and its integration into a primarily computer graphics system. Although perhaps difficult to generalize from this particular experience (a case study if you will), nevertheless there are two points that are considered key to successful installation and usage of the device:

- Signal routing
- Synchronization.

Signal routing refers to proper signal input and output, namely concerning video feeds (input to the eye tracker) and (typically rs232) serial data capture and interpretation (output from the eye tracker). Synchronization refers to the proper coordinate mapping of the eye tracker's reference frame to the application responsible for generating the visual stimulus that will be seen by the observer. Proper mapping will ensure correct registration of both eye tracker calibration and subsequent data analysis. The actual data mapping is carried out in software (see the next chapter), however, the hardware component that facilitates proper software mapping is the eye tracker's capability of measuring the application's display extents in its own reference frame. This is usually provided by the eye tracker if it allows manual positioning and readout of its calibration cursor.

There are two primary sources where further information can be obtained on system installation and setup. First, of course, is the manufacturer's manual that will typically be included with the eye tracking device. Second, the best resource for installation and usage of the equipment is from the users themselves. Users of eye trackers typically report, in a somewhat formal way, what they use and how they use it in most technical papers on eye tracking research. These can be found in various journal articles, such as *Vision Research, Behavior Research Methods, Instruments, and Computers (BRMIC)*, and conference proceedings. There are various conferences that deal with eye tracking, either directly or indirectly. For example, conferences that deal with computer graphics (e.g., SIGGRAPH, EuroGraphics, or Graphics Interface), human–computer interaction (e.g., SIGCHI), or virtual reality (e.g., VRST), may include papers that discuss the use of eye trackers, but their apparatus description may be somewhat indirect insofar as the objective of the report typically does not deal with the eye tracker itself, rather it concentrates more on the results obtained through its application. Examples of such applications are given in Part III of this text. At this time, there are two main conferences that deal more directly with eye tracking: the European Conference on Eye Movements (ECEM), and the U.S.-based Eye Tracking Research & Applications (ETRA). Finally, the eye movement email listservs *eye-movements* and *eyemov-l* are excellent on-line "gathering places" of eye tracker researchers. Here, questions on eye trackers may be answered directly (and usually promptly) by users of similar equipment or even from other vendors.

7

Head-Mounted System Software Development

In designing a graphical eye tracking application, the most important requirement is mapping of eye tracker coordinates to the appropriate application program's reference frame. The eye tracker calculates the viewer's Point Of Regard (POR) relative to the eye tracker's screen reference frame, e.g., a 512×512 pixel plane, perpendicular to the optical axis. That is, for each eye, the eye tracker returns a sample coordinate pair of x- and y-coordinates of the POR at each sampling cycle (e.g., once every $\sim$16 ms for a 60 Hz device). This coordinate pair must be mapped to the extents of the application program's viewing window.

If a binocular eye tracker is used, two coordinate sample pairs are returned during each sampling cycle, x_l, y_l for the left POR and x_r, y_r for the right eye. A Virtual Reality (VR) application must, in the same update cycle, also map the coordinates of the head's position and orientation tracking device (e.g., typically a 6 DOF tracker is used).

The following sections discuss the mapping of eye tracker screen coordinates to application program coordinates for both monocular and binocular applications. In the monocular case, a typical 2D image-viewing application is expected, and the coordinates are mapped to the 2D (orthogonal) viewport coordinates accordingly (the viewport coordinates are expected to match the dimensions of the image being displayed). In the binocular (VR) case, the eye tracker coordinates are mapped to the dimensions of the near viewing plane of the viewing frustum. For both calculations, a method of measuring the application's viewport dimensions in the eye tracker's reference frame is described, where the eye tracker's own (fine-resolution) cursor control is used to obtain the measurements. This information should be sufficient for most 2D image viewing applications.

For VR applications, subsequent sections describe the required mapping of the head position/orientation tracking device. Although this section should generalize to most kinds of 6 DOF tracking devices, the discussion in some cases is specific to the Ascension Flock Of Birds (FOB) d.c. electromagnetic 6 DOF tracker. This section discusses how to obtain the head-centric view and vectors from the matrix returned by the tracker, and also explains the transformation of an arbitrary vector using the obtained transformation matrix.[1] This latter derivation is used to transform the gaze vector to head-centric coordinates, which is initially obtained from, and relative to, the binocular eye tracker's left and right POR measurement.

Finally, a derivation of the calculation of the gaze vector in 3D is given, and a method is suggested for calculating the three-dimensional gaze point in VR.

7.1 Mapping Eye Tracker Screen Coordinates

When working with the eye tracker, the data obtained from the tracker must be mapped to a range appropriate to the given application. If working in VR, the 2D eye tracker data, expressed in eye tracker screen coordinates, must be mapped to the 2D dimensions of the near viewing frustum. If working with images, the 2D eye tracker data must be mapped to the 2D display image coordinates.

In general, if $x' \in [a,b]$ needs to be mapped to the range $[c,d]$, we have:

$$x = c + \frac{(x'-a)(d-c)}{(b-a)}. \tag{7.1}$$

This is a linear mapping between two (one-dimensional) coordinate systems (or lines in this case). Equation (7.1) has a straightforward interpretation:

1. The value x' is translated (shifted) to its origin, by subtracting a.
2. The value $(x'-a)$ is then normalized by dividing through by the range $(b-a)$.
3. The normalized value $(x'-a)/(b-a)$ is then scaled to the new range $(d-c)$.
4. Finally, the new value is translated (shifted) to its proper relative location in the new range by adding c.

7.1.1 Mapping Screen Coordinates to the 3D Viewing Frustum

The 3D viewing frustum employed in the perspective viewing transformations is defined by the parameters left, right, bottom, top, near, far, e.g., as used in the OpenGL function call glFrustum(). Figure 7.1 shows the dimensions of the eye tracker screen (left) and the dimensions of the viewing frustum (right). Note that the eye tracker origin is the top-left of the screen and the viewing frustum's origin is bottom-left (this is a common discrepancy between imaging and graphics). To convert the eye tracker coordinates (x',y') to graphics coordinates (x,y), using Equation (7.1), we have:

---

[1] Equivalently, the rotation quaternion may be used, if these data are available from the head tracker.

$$x = \text{left} + \frac{x'(\text{right} - \text{left})}{512} \tag{7.2}$$

$$y = \text{bottom} + \frac{(512 - y')(\text{top} - \text{bottom})}{512}. \tag{7.3}$$

Note that the term $(512 - y')$ in Equation (7.3) handles the y-coordinate mirror transformation so that the top-left origin of the eye tracker screen is converted to the bottom-left of the viewing frustum.

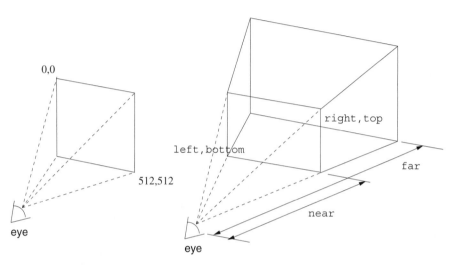

Fig. 7.1. Eye tracker to VR mapping.

If typical dimensions of the near plane of the graphics viewing frustum are 640 × 480, with the origin at $(0,0)$, then Equations (7.2) and (7.3) reduce to:

$$x = \frac{x'(640)}{512} \quad = \quad x'(1.3) \tag{7.4}$$

$$y = \frac{(512 - y')(480)}{512} \quad = \quad (512 - y')(.9375). \tag{7.5}$$

7.1.2 Mapping Screen Coordinates to the 2D Image

The linear mapping between eye tracker screen coordinates and 2D image coordinates is handled similarly. If the image dimensions are 640 × 480, then Equations (7.4) and (7.5) are used without change. Note that an image with dimensions 600 × 450 appears to fit the display of the TV in the Clemson VRET lab better.[2] In this case, Equations (7.4) and (7.5) become:

---

[2] A few of the pixels of the image do not fit on the TV display, possibly due to the NTSC flicker-free filter used to encode the SGI console video signal.

$$x = \frac{x'(600)}{512} = x'(1.171875) \tag{7.6}$$

$$y = \frac{(512 - y')(450)}{512} = (512 - y')(.87890625). \tag{7.7}$$

7.1.3 Measuring Eye Tracker Screen Coordinate Extents

The above coordinate mapping procedures assume that the eye tracker coordinates are in the range $[0, 511]$. In reality, the *usable*, or effective coordinates will be dependent on: (a) the size of the application window, and (b) the position of the application window. For example, if an image display program runs wherein an image is displayed in a 600×450 window, and this window is positioned arbitrarily on the graphics console, then the eye tracking coordinates of interest are restricted only to the area covered by the application window.

The following example illustrates the eye tracker/application program coordinate mapping problem using a schematic depicting the scene imaging display produced by an eye tracker manufactured by ISCAN. The ISCAN eye tracker is a fairly popular video-based corneal-reflection eye tracker which provides an image of the scene viewed by the subject and is seen by the operator on small ISCAN black and white scene monitors. Most eye trackers provide this kind of functionality. Because the application program window is not likely to cover the entire scene display (as seen in the scene monitors), only a restricted range of eye tracker coordinates will be of relevance to the eye tracking experiment. To use the ISCAN as an example, consider a 600×450 image display application. Once the window is positioned so that it fully covers the viewing display, it may appear slightly off-center on the ISCAN black-/white monitor, as sketched in Figure 7.2.

The trick to the calculation of the proper mapping between eye tracker and application coordinates is the measurement of the application window's extents, in the eye tracker's reference frame. This is accomplished by using the eye tracker itself. One of the eye tracker's most useful features, if available for this purpose, is its ability to move its crosshair cursor. The software may allow cursor fine (pixel by pixel) or coarse (in jumps of ten pixels) movement. Furthermore, a data screen at the bottom of the scene monitor may indicate the coordinates of the cursor, relative to the eye tracker's reference frame.

To obtain the extents of the application window in the eye tracker's reference frame, simply move the cursor to the corners of the application window. Use these coordinates in the above mapping formulas. The following measurement "recipe" should provide an almost exact mapping, and only needs to be performed once. Assuming the application window's dimensions are fixed, the mapping obtained from this procedure can be hardcoded into the application. Here are the steps:

1. Position your display window so that it covers the display fully, e.g., in the case of the image stimulus, the window should just cover the entire viewing (e.g., TV or virtual reality Head-Mounted Display (HMD)) display.
2. Use the eye tracker's cursor positioning utility to measure the viewing area's extents (toggle between course and fine cursor movement).
3. Calculate the mapping.
4. In Reset mode, position the eye tracker cursor in the middle of the display.

An important consequence of the above is that following the mapping calculation, it should be possible to always position the application window in the same place, provided that the program displays the calibration point obtained from the eye tracker mapped to local image coordinates. When the application starts up (and the eye tracker is on and in Reset mode), simply position the application so that the central calibration points (displayed when the tracker is in Reset mode) line up.

To conclude this example, assume that if the ISCAN cursor is moved to the corners of the drawable application area, the measurements would appear as shown in Figure 7.2. Based on those measurements, the mapping is:

$$x = \frac{x' - 51}{(482 - 51 + 1)}(600) \tag{7.8}$$

$$y = 449 - \frac{y' - 53}{(446 - 53 + 1)}(450). \tag{7.9}$$

The central point on the ISCAN display is $(267, 250)$. Note that y is subtracted from 449 to take care of the image/graphics vertical origin flip.

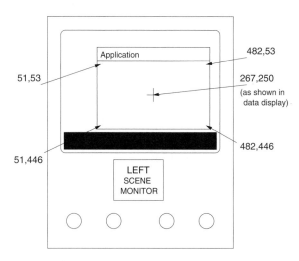

Fig. 7.2. Example mapping measurement.

7.2 Mapping Flock Of Birds Tracker Coordinates

In virtual reality, the position and orientation of the head is typically delivered by
a real-time head tracker. In our case, we have a flock of birds d.c. electromagnetic
tracker from Ascension. The tracker reports 6 degree of freedom information regarding sensor position and orientation. The latter is given in terms of Euler angles. Euler
angles determine the orientation of a vector in three-space by specifying the required
rotations of the origin's coordinate axes. These angles are known by several names,
but in essence each describes a particular rotation angle about one of the principal
axes. Common names describing Euler angles are given in Table 7.1 and the geometry is sketched in Figure 7.3, where roll, pitch, and yaw angles are represented
by R, E, and A, respectively. Each of these rotations is represented by the familiar
homogeneous rotation matrices:

$$
\text{Roll (rot z)} \quad = \quad \mathbf{R}_z =
\begin{bmatrix}
\cos R & \sin R & 0 & 0 \\
-\sin R & \cos R & 0 & 0 \\
0 & 0 & 1 & 0 \\
0 & 0 & 0 & 1
\end{bmatrix}
$$

$$
\text{Pitch (rot x)} \quad = \quad \mathbf{R}_x =
\begin{bmatrix}
1 & 0 & 0 & 0 \\
0 & \cos E & \sin E & 0 \\
0 & -\sin E & \cos E & 0 \\
0 & 0 & 0 & 1
\end{bmatrix}
$$

$$
\text{Yaw (rot y)} \quad = \quad \mathbf{R}_y =
\begin{bmatrix}
\cos A & 0 & -\sin A & 0 \\
0 & 1 & 0 & 0 \\
\sin A & 0 & \cos A & 0 \\
0 & 0 & 0 & 1
\end{bmatrix}.
$$

The composite 4×4 matrix, containing all of the above transformations rolled into
one, is:

$$
\mathbf{M} \;=\; \mathbf{R}_z \mathbf{R}_x \mathbf{R}_y \;=
$$

$$
\begin{bmatrix}
\cos R \cos A + \sin R \sin E \sin A & \sin R \cos E & -\cos R \sin A + \sin R \sin E \cos A & 0 \\
-\sin R \cos A + \cos R \sin E \sin A & \cos R \cos E & \sin R \sin A + \cos R \sin E \cos A & 0 \\
\cos E \sin A & -\sin E & \cos E \cos A & 0 \\
0 & 0 & 0 & 1
\end{bmatrix}.
$$

The FOB delivers a similar 4×4 matrix,

$$
\mathbf{F} \;=
$$

$$
\begin{bmatrix}
\cos E \cos A & \cos E \sin A & -\sin E & 0 \\
-\cos R \sin A + \sin R \sin E \cos A & \cos R \cos A + \sin R \sin E \sin A & \sin R \cos E & 0 \\
\sin R \sin A + \cos R \sin E \cos A & -\sin R \cos A + \cos R \sin E \sin A & \cos R \cos E & 0 \\
0 & 0 & 0 & 1
\end{bmatrix},
$$

where the matrix elements are slightly rearranged, such that

$$
\mathbf{M}[i, j] = \mathbf{F}[(i+1)\%3][(j+1)\%3],
$$

e.g., row 1 of matrix **M** is now row 2 in matrix **F**, row 2 is now row 3, and row 3 is now row 1. Columns are interchanged similarly, where column 1 of matrix **M** is now column 2 in matrix **F**, column 2 is now column 3, and column 3 is now column 1. This "off-by-one" shift present in the Bird matrix may be due to the non-C style indexing which starts at 1 instead of 0.

Table 7.1. Euler angles.

rot y	rot x	rot z
Yaw	Pitch	Roll
Azimuth	Elevation	Roll
Longitude	Latitude	Roll

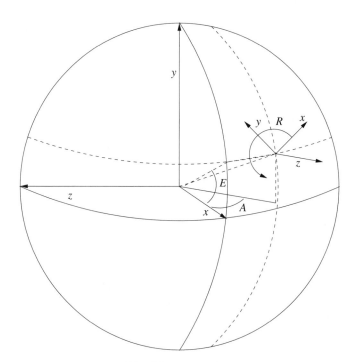

Fig. 7.3. Euler angles.

7.2.1 Obtaining the Transformed View Vector

To calculate the transformed view vector $\mathbf{v}'$, assume the initial view vector is $\mathbf{v} = (0, 0, -1)$ (looking down the z-axis), and apply the composite transformation (in homogeneous coordinates):[3]

$$\mathbf{vM} = \begin{bmatrix} 0 & 0 & -1 & 1 \end{bmatrix} *$$

$$\begin{bmatrix} \cos R \cos A + \sin R \sin E \sin A & \sin R \cos E & -\cos R \sin A + \sin R \sin E \cos A & 0 \\ -\sin R \cos A + \cos R \sin E \sin A & \cos R \cos E & \sin R \sin A + \cos R \sin E \cos A & 0 \\ \cos E \sin A & -\sin E & \cos E \cos A & 0 \\ 0 & 0 & 0 & 1 \end{bmatrix}$$

$$= \begin{bmatrix} -\cos E \sin A & \sin E & \cos E \cos A & 1 \end{bmatrix},$$

which is simply the third row of $\mathbf{M}$, negated. By inspection of the Bird matrix $\mathbf{F}$, the view vector is simply obtained by selecting appropriate entries in the matrix; i.e.,

$$\mathbf{v}' = \begin{bmatrix} -\mathbf{F}[0,1] & -\mathbf{F}[0,2] & \mathbf{F}[0,0] & 1 \end{bmatrix}. \tag{7.10}$$

7.2.2 Obtaining the Transformed Up Vector

The transformed up vector is obtained in a similar manner to the transformed view vector. To calculate the transformed up vector $\mathbf{u}'$, assume the initial up vector is $\mathbf{u} = (0,1,0)$ (looking up the y-axis), and apply the composite transformation (in homogeneous coordinates):

$$\mathbf{uM} = \begin{bmatrix} 0 & 1 & 0 & 1 \end{bmatrix} *$$

$$\begin{bmatrix} \cos R \cos A + \sin R \sin E \sin A & \sin R \cos E & -\cos R \sin A + \sin R \sin E \cos A & 0 \\ -\sin R \cos A + \cos R \sin E \sin A & \cos R \cos E & \sin R \sin A + \cos R \sin E \cos A & 0 \\ \cos E \sin A & -\sin E & \cos E \cos A & 0 \\ 0 & 0 & 0 & 1 \end{bmatrix}$$

$$= \begin{bmatrix} -\sin R \cos A + \cos R \sin E \sin A & \cos R \cos E & \sin R \sin A + \cos R \sin E \cos A & 1 \end{bmatrix}.$$

By inspection of the Bird matrix $\mathbf{F}$, the transformed up vector is simply obtained by selecting appropriate entries in the matrix; i.e.,

$$\mathbf{u}' = \begin{bmatrix} \mathbf{F}[2,1] & \mathbf{F}[2,2] & \mathbf{F}[2,0] & 1 \end{bmatrix}. \tag{7.11}$$

Because of our setup in the lab, the z-axis is on the opposite side of the Bird transmitter (behind the Bird emblem on the transmitter). For this reason, the z-component of the up vector is negated, i.e.,

$$\mathbf{u}' = \begin{bmatrix} \mathbf{F}[2,1] & \mathbf{F}[2,2] & -\mathbf{F}[2,0] & 1 \end{bmatrix}. \tag{7.12}$$

Note that the negation of the z-component of the transformed view vector does not make a difference because the term is a product of cosines.

---

[3] Recall trigonometric identities: $\sin(-\theta) = -\sin(\theta)$ and $\cos(-\theta) = \cos(\theta)$.

7.2.3 Transforming an Arbitrary Vector

To transform an arbitrary vector, an operation similar to the transformations of the up and view vectors is performed. To calculate the transformed arbitrary vector, $\mathbf{w} = \begin{bmatrix} x & y & z & 1 \end{bmatrix}$, apply the composite transformation by multiplying by the transformation matrix $\mathbf{M}$ (in homogeneous coordinates):

$$\mathbf{wM} = \begin{bmatrix} x & y & z & 1 \end{bmatrix} *$$
$$\begin{bmatrix} \cos R \cos A + \sin R \sin E \sin A & \sin R \cos E & -\cos R \sin A + \sin R \sin E \cos A & 0 \\ -\sin R \cos A + \cos R \sin E \sin A & \cos R \cos E & \sin R \sin A + \cos R \sin E \cos A & 0 \\ \cos E \sin A & -\sin E & \cos E \cos A & 0 \\ 0 & 0 & 0 & 1 \end{bmatrix}$$

which gives transformed vector $\mathbf{w}'$,

$$\mathbf{w}' = \begin{bmatrix} x\mathbf{M}[1,1] + y\mathbf{M}[2,1] + z\mathbf{M}[3,1] \\ x\mathbf{M}[1,2] + y\mathbf{M}[2,2] + z\mathbf{M}[3,2] \\ x\mathbf{M}[1,3] + y\mathbf{M}[2,3] + z\mathbf{M}[3,3] \\ 1 \end{bmatrix}^T.$$

To use the Bird matrix, there is unfortunately no simple way to select the appropriate matrix elements to directly obtain $\mathbf{w}'$. Probably the best bet would be to undo the "off-by-one" shift present in the Bird matrix. On the other hand, hardcoding the solution may be the fastest method. This rather inelegant, but luckily localized, operation looks like this:

$$\mathbf{w}' = \begin{bmatrix} x\mathbf{F}[1,1] + y\mathbf{F}[2,1] + z\mathbf{F}[0,1] \\ x\mathbf{F}[1,2] + y\mathbf{F}[2,2] + z\mathbf{F}[0,2] \\ x\mathbf{F}[1,0] + y\mathbf{F}[2,0] + z\mathbf{F}[0,0] \\ 1 \end{bmatrix}^T. \tag{7.13}$$

Furthermore, as in the up vector transformation, it appears that the negation of the z component may also be necessary. If so, the above equation will need to be rewritten as

$$\mathbf{w}' = \begin{bmatrix} x\mathbf{F}[1,1] + y\mathbf{F}[2,1] + z\mathbf{F}[0,1] \\ x\mathbf{F}[1,2] + y\mathbf{F}[2,2] + z\mathbf{F}[0,2] \\ -(x\mathbf{F}[1,0] + y\mathbf{F}[2,0] + z\mathbf{F}[0,0]) \\ 1 \end{bmatrix}^T. \tag{7.14}$$

7.3 3D Gaze Point Calculation

The calculation of the gaze point in three-space depends only on the relative positions of the two eyes in the horizontal axis. The parameters of interest here are the three-dimensional virtual coordinates of the gaze point, (x_g, y_g, z_g), which can be determined from traditional stereo geometry calculations. Figure 7.4 illustrates the basic binocular geometry. Helmet tracking determines helmet position and orthogonal directional and up vectors, which determine viewer-local coordinates shown in the diagram. The helmet position is the origin (x_h, y_h, z_h), the helmet directional vector is the optical (viewer-local) z-axis, and the helmet up vector is the viewer-local y-axis.

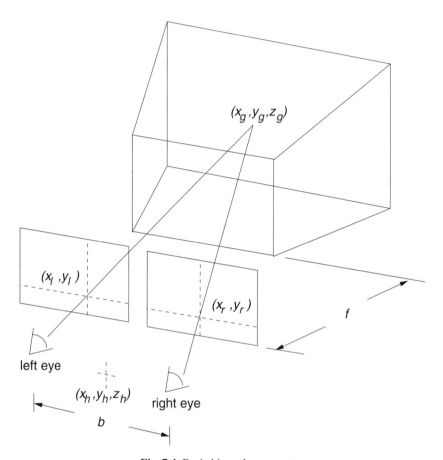

Fig. 7.4. Basic binocular geometry.

Given instantaneous, eye tracked, viewer-local coordinates (mapped from eye tracker screen coordinates to the near view plane coordinates), (x_l, y_l) and (x_r, y_r) in the left and right view planes, at focal distance f along the viewer-local z-axis, we can determine viewer-local coordinates of the gaze point (x_g, y_g, z_g) by deriving the stereo equations parametrically. First, express both the left and right view lines in terms of the linear parameter s. These lines originate at the eye centers $(x_h - b/2, y_h, z_h)$ and $(x_h + b/2, y_h, z_h)$ and pass through (x_l, y_l, f) and (x_r, y_r, f), respectively. The left view line is (in vector form):

$$\left[(1-s)(x_h-b/2)+sx_l \ (1-s)y_h+sy_l \ (1-s)z_h+sf\right] \tag{7.15}$$

and the right view line is (in vector form):

$$\left[(1-s)(x_h+b/2)+sx_r \ (1-s)y_h+sy_r \ (1-s)z_h+sf\right], \tag{7.16}$$

where b is the disparity distance between the left and right eye centers, and f is the distance to the near viewing plane. To find the central view line originating at the local center (x_h,y_h,z_h), calculate the intersection of the left and right view lines by solving for s using the x-coordinates of both lines, as given in Equations (7.15) and (7.16):

$$(1-s)(x_h-b/2)+sx_l = (1-s)(x_h+b/2)+sx_r$$
$$(x_h-b/2)-s(x_h-b/2)+sx_l = (x_h+b/2)-s(x_h+b/2)+sx_r$$
$$s(x_l-x_r+b) = b$$

$$s = \frac{b}{x_l-x_r+b}. \tag{7.17}$$

The interpolant s is then used in the parametric equation of the central view line, to give the gaze point at the intersection of both view lines:

$$x_g = (1-s)x_h+s((x_l+x_r)/2)$$
$$y_g = (1-s)y_h+s((y_l+y_r)/2)$$
$$z_g = (1-s)z_h+sf,$$

giving:

$$x_g = \left(1-\frac{b}{x_l-x_r+b}\right)x_h+\left(\frac{b}{x_l-x_r+b}\right)\left(\frac{x_l+x_r}{2}\right) \tag{7.18}$$

$$y_g = \left(1-\frac{b}{x_l-x_r+b}\right)y_h+\left(\frac{b}{x_l-x_r+b}\right)\left(\frac{y_l+y_r}{2}\right) \tag{7.19}$$

$$z_g = \left(1-\frac{b}{x_l-x_r+b}\right)z_h+\left(\frac{b}{x_l-x_r+b}\right)f. \tag{7.20}$$

Eye positions (at viewer local coordinates $(x_h-b/2,y_h,z_h)$ and $(x_h+b/2,y_h,z_h)$), the gaze point, and an up vector orthogonal to the plane of the three points then determine the view volume appropriate for display to each eye screen.

The gaze point, as defined above, is given by the addition of a scaled offset to the view vector originally defined by the helmet position and central view line in virtual world coordinates.[4] The gaze point can be expressed parametrically as a point on a

---

[4] Note that the vertical eye tracked coordinates y_l and y_r are expected to be equal (because gaze coordinates are assumed to be epipolar); the vertical coordinate of the central view vector defined by $(y_l+y_r)/2$ is somewhat extraneous; either y_l or y_r would do for the calculation of the gaze vector. However, because eye tracker data are also expected to be noisy, this averaging of the vertical coordinates enforces the epipolar assumption.

ray with origin (x_h, y_h, z_h) and the helmet position, with the ray emanating along a vector scaled by parameter s. That is, rewriting Equations (7.18) through (7.20), we have:

$$x_g = x_h + s\left(\frac{x_l + x_r}{2} - x_h\right)$$

$$y_g = y_h + s\left(\frac{y_l + y_r}{2} - y_h\right)$$

$$z_g = z_h + s\left(f - z_h\right)$$

or, in vector notation,

$$\mathbf{g} = \mathbf{h} + s\mathbf{v}, \tag{7.21}$$

where $\mathbf{h}$ is the head position, $\mathbf{v}$ is the central view vector, and s is the scale parameter as defined in Equation (7.17). Note that the view vector used here is not related to the view vector given by the head tracker. It should be noted that the view vector $\mathbf{v}$ is obtained by subtracting the helmet position from the midpoint of the eye tracked x-coordinate and focal distance to the near view plane; i.e.,

$$\mathbf{v} = \begin{bmatrix} (x_l + x_r)/2 \\ (y_l + y_r)/2 \\ f \end{bmatrix} - \begin{bmatrix} x_h \\ y_h \\ z_h \end{bmatrix}$$

$$= \mathbf{m} - \mathbf{h},$$

where $\mathbf{m}$ denotes the left and right eye coordinate midpoint. To transform the vector $\mathbf{v}$ to the proper (instantaneous) head orientation, this vector should be normalized, then multiplied by the orientation matrix returned by the head tracker (see Section 7.2 in general and Section 7.2.3 in particular). This new vector, call it $\mathbf{m'}$, should be substituted for $\mathbf{m}$ above to define $\mathbf{v}$ for use in Equation (7.21); i.e.,

$$\mathbf{g} = \mathbf{h} + s(\mathbf{m'} - \mathbf{h}). \tag{7.22}$$

7.3.1 Parametric Ray Representation of Gaze Direction

Equation (7.22) gives the coordinates of the gaze point through a parametric representation (e.g., a point along a line) such that the depth of the three-dimensional point of regard in world coordinates is valid only if $s > 0$. Given the gaze point $\mathbf{g}$ and the location of the helmet $\mathbf{h}$, we can obtain just the three-dimensional gaze vector v that specifies the direction of gaze (but not the actual fixation point). This direction vector is given by:

$$v = \mathbf{g} - \mathbf{h} \tag{7.23}$$

$$= (\mathbf{h} + s\mathbf{v}) - \mathbf{h} \tag{7.24}$$

$$= s\mathbf{v} \tag{7.25}$$

$$= \left(\frac{b}{x_l - x_r + b}\right) \begin{bmatrix} (x_l + x_r)/2 - x_h \\ (y_l + y_r)/2 - y_h \\ (f - z_h) \end{bmatrix}, \tag{7.26}$$

where $\mathbf{v}$ is defined as either $\mathbf{m} - \mathbf{h}$ as before, or as $\mathbf{m}' - \mathbf{h}$ as in Equation (7.22). Given the helmet position $\mathbf{h}$ and the gaze direction $\mathbf{v}$, we can express the gaze direction via a parametric representation of a ray using a linear interpolant t:

$$\text{gaze}(t) = \mathbf{h} + t\mathbf{v}, \quad t > 0, \tag{7.27}$$

where h is the ray's origin (a point; the helmet position), and $\mathbf{v}$ is the ray direction (the gaze vector). (Note that adding h to $t\mathbf{v}$ results in the original expression of the gaze point $\mathbf{g}$ given by Equation (7.21), provided $t = 1$.) The formulation of the gaze direction given by Equation (7.27) can then be used for testing virtual fixation coordinates via traditional ray/polygon intersection calculations commonly used in ray-tracing.

7.4 Virtual Gaze Intersection Point Coordinates

In 3D eye tracking studies, we are often interested in knowing the location of one's gaze, or more importantly one's fixation, relative to some feature in the scene. In VR applications, we'd like to calculate the fixation location in the virtual world and thus identify the object of interest. The identification of the object of interest can be accomplished following traditional ray/polygon intersection calculations, as employed in ray-tracing (Glassner, 1989).

The fixated object of interest is the one closest to the viewer that intersects the gaze ray. This object is found by testing all polygons in the scene for intersection with the gaze ray. The polygon closest to the viewer is then assumed to be the one fixated by the viewer (assuming all polygons in the scene are opaque).

7.4.1 Ray/Plane Intersection

The calculation of an intersection between a ray and all polygons in the scene is usually obtained via a parametric representation of the ray; e.g.,

$$\text{ray}(t) = r_o + t\mathbf{r_d}, \tag{7.28}$$

where r_o defines the ray's origin (a point), and $\mathbf{r_d}$ defines the ray direction (a vector). Note the similarity between Equations (7.28) and (7.27); there, h is the head position and $\mathbf{v}$ is the gaze direction. To find the intersection of the ray with a polygon, calculate the interpolant value where the ray intersects each polygon, and examine all the intersections where $t > 0$. If $t < 0$, the object may intersect the ray, but behind the viewer.

Recall the plane equation $Ax + By + Cz + D = 0$, where $A^2 + B^2 + C^2 = 1$; i.e., A, B, and C define the plane normal. To obtain the ray/polygon intersection, substitute Equation (7.28) into the plane equation:

$$A(x_o + tx_d) + B(x_o + ty_d) + C(z_o + tz_d) + D = 0 \qquad (7.29)$$

and solve for t:

$$t = -\frac{Ax_o + Bx_o + Cz_o + D}{Ax_d + By_d + Cz_d} \qquad (7.30)$$

or, in vector notation:

$$t = \frac{-(\mathbf{N} \cdot r_o + D)}{\mathbf{N} \cdot \mathbf{r_d}}. \qquad (7.31)$$

A few observations of the above simplify the implementation.

1. Here $\mathbf{N}$, the face normal, is really $-\mathbf{N}$, because what we're doing is calculating the angle between the ray and face normal. To get the angle, we need both ray and normal to be pointing in the same relative direction. This situation is depicted in Figure 7.5.
2. In Equation (7.31), the denominator will cause problems in the implementation should it evaluate to 0. However, if the denominator is 0 (i.e., if $\mathbf{N} \cdot \mathbf{r_d} = 0$), then the cosine between the vectors is 0, which means that the angle between the two vectors is 90° which means the ray and plane are parallel and don't intersect. Thus, to avoid dividing by zero, and to speed up the computation, evaluate the denominator first. If it is sufficiently close to zero, don't evaluate the intersection further; we know the ray and polygon will not intersect.
3. Point 2 above can be further exploited by noting that if the dot product is greater than 0, then the surface is hidden to the viewer.

The first part of the intersection algorithm follows from the above and is given in Listing 7.1.

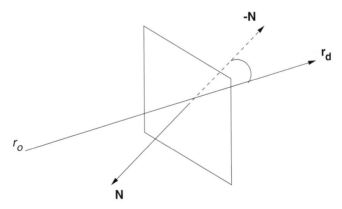

Fig. 7.5. Ray/plane geometry.

$$
\begin{aligned}
&v_d = \mathbf{N} \cdot \mathbf{r_d}; &&\text{// denominator}\\
&\textbf{if } (v_d < 0)\ \{\\
&\quad v_o = -(\mathbf{N} \cdot r_o + D); &&\text{// numerator}\\
&\quad t = v_o / v_d;\\
&\}
\end{aligned}
$$

Listing 7.1. Ray/polygon intersection.

In the algorithm, the intersection parameter t defines the point of intersection along the ray at the plane defined by the normal $\mathbf{N}$. That is, if $t > 0$, then the point of intersection p is given by:

$$p = r_o + t\mathbf{r_d}. \tag{7.32}$$

Note that the above only gives the intersection point of the ray and the (infinite!) plane defined by the polygon's face normal. Because the normal defines a plane of infinite extent, we need to test the point p to see if it lies within the confines of the polygonal region, which is defined by the polygon's edges. This is essentially the "point-in-polygon" problem.

7.4.2 Point-In-Polygon Problem

To test whether a point p lies inside a polygon (defined by its plane equation which specifies a plane of finite extent), we need to test the point against all edges of the polygon. To do this, the following algorithm is used.

For each edge:
1. Get the plane perpendicular to the face normal $\mathbf{N}$, which passes through the edge's two vertices A and B. The perpendicular face normal $\mathbf{N}'$ is obtained by calculating the cross product of the original face normal with the edge; i.e.,
$$\mathbf{N}' = \mathbf{N} \times (B - A),$$
where the face vertices A and B are specified in counterclockwise order. This is shown in Figure 7.6.
2. Get the perpendicular plane's equation by calculating D using either of A or B as the point on the plane (e.g., plug in either A or B into the plane equation; then solve for D).
3. Test point p to see if it lies "above" or "below" the perpendicular plane. This is done by plugging p into the perpendicular plane's equation and testing the result. If the result is greater than 0, then p is "above" the plane.
4. If p is "above" all perpendicular planes, as defined by successive pairs of vertices, then the point is "boxed in" by all the polygon's edges, and so must lie inside the polygon as originally defined by its face normal $\mathbf{N}$.

Note that this is just one solution to the point-in-polygon problem; other, possibly faster, algorithms may be available.

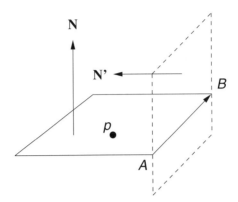

Fig. 7.6. Point-in-polygon geometry.

The calculated intersection points p, termed here Gaze Intersection Points (GIPs) for brevity, are each found on the closest polygon to the viewer intersecting the gaze ray, assuming all polygons are opaque. The resulting ray-casting algorithm generates a scanpath constrained to lie on polygonal regions within the virtual environment. Eye movement analysis is required to identify fixation points in the environment (see Chapter 12).

7.5 Data Representation and Storage

Data collection is straightforward. For 2D imaging applications, the point of regard can be recorded, along with the timestamp of each sample. Listing 7.2 shows a pseu-docode sample data structure suitable for recording the point of regard. Because the number of samples may be rather large, depending on sampling rate and duration of viewing trial, a linked list may be used to dynamically record the data. In Listing 7.2, the data type queue denotes a linked list pointer.

For 3D VR applications, the simplest data structure to record the calculated gaze point is similar to the 2D structure, with the addition of the third z component. Data captured (i.e., the three-dimensional gaze point) can then be displayed for review of the session statically in the VR environment in which it was recorded, as shown in Figure 7.7. In Figure 7.7 consecutive gaze points have been joined by straight line segments to display a three-dimensional scanpath. Note that in this visualization it may be easy to misinterpret the scanpath data with the position of the head. To disambiguate the two, head position (and tilt angle) may also be stored/displayed.

```
typedef struct _PORNode {
  _PORNode* link ;                    // linked list node
  int      x,y;                       // POR data
  double   t;                         // timestamp (usually in ms)
} PORnode;
```

Listing 7.2. 2D imaging point of regard data structure.

Fig. 7.7. Example of three-dimensional gaze point captured in VR. Courtesy of Tom Auchter and Jeremy Barron. Reproduced with permission, Clemson University.

There is one fairly important aspect of data storage which is worth mentioning, namely, proper labeling of the data. Because eye movement experiments tend to generate voluminous amounts of data, it is imperative to properly label data files. Files should encode the number of the subject viewing the imagery, the imagery itself, the trial number, and the state of the data (raw or processed). One of the easiest methods for encoding this information is in the data filename itself (at least on file systems that allow long filenames).

7.6 Summary and Further Reading

This chapter focused on the main programming aspects of proper collection of raw eye movement data from the eye tracker, including appropriate mapping of raw eye movement coordinates. Coordinate mapping is crucial for both eye tracker calibration and subsequent eye movement coordination registration in the application reference frame.

Two important programming components have been omitted from this chapter, namely, a mechanism for user interaction and for stimulus display. Generally, stimulus display requires knowledge of some sort of graphical Application Program Interface (API) that enables display of either digital images or graphics primitives (as required for VR). A particularly suitable API for both purposes is OpenGL (see <http://www.opengl.org>). For user interface development, particularly if developing on a UNIX or Linux platform, either the GTK Graphical User Interface (GUI) or Qt API is recommended. GTK is the freely available GNU Toolkit (see <http://www.gtk.org>). Qt is available from <http://trolltech.com>.

8

Head-Mounted System Calibration

Currently, most video-based eye trackers require calibration. This is usually a sequence of simple stimuli displayed sequentially at far extents of the viewing region. The eye tracker calculates the Point Of Regard (POR) by measuring the relative observed position of the pupil and corneal reflection at these locations, and then (most likely) interpolates the POR value at intermediate eye positions. The individual stimuli used for this purpose are simple white dots or crosshairs on a black background. In cases where the eye tracker is used in an outside setting (e.g., for use during driving or while walking outside the lab), calibration marks may be made from simple targets such as tape or other visible markers fixed to objects in the environment. The purpose of calibration is to present a sequence of visible points at fairly extreme viewing angle ranges (e.g., upper-left, upper-right, lower-left, lower-right). These extrema points should be chosen to provide a sufficiently large enough coordinate range to allow the eye tracker to interpolate the viewer's POR between extrema points. Most (video-based) eye trackers provide built-in calibration techniques where a number of such extrema points (e.g., three, five, or nine typically) are presented in order.

Apart from selection and/or presentation of properly distributed calibration stimulus points, a secondary but equally important goal of calibrating the eye tracker is correct adjustment of the eye tracker's optics and threshold levels to allow the device to properly recognize critical visual elements of the eyes. In the case of a video-based corneal reflection eye tracker, the device attempts to automatically detect the eye's pupil center and the center of the corneal reflection(s) (see Section 5.4). Corresponding to these elements, the eye tracker software may offer a mechanism to adjust both pupil and corneal reflection detection thresholds. Both must be set so that the eye tracker can easily detect the centers of the pupil and the corneal reflection, without either losing detection of either or confusing these targets with distracting artifacts such as eyelashes, rims of glasses, or contact lenses, to name a few usual problematic features. Figure 8.1 shows the eye images as seen by the eye tracker. Notice that the eye tracker has correctly labeled (with black crosshairs) the pupil center and one of two[1] corneal reflections while the subject was viewing each of five calibration targets. Figure 8.1(a) shows a fairly good eye image with properly identified pupil and corneal reflections; notice also the pupil threshold "bleeding" that is seen on the sclera to the bottom and left of the pupil. The pupil threshold may have been set a touch lower to eliminate this artifact.

Figures 8.1d and e show potential problems caused by long eyelashes. Eyelashes (particularly with heavy application of mascara), contact lenses, and eyeglasses may pose problems by providing reflecting surfaces in the scene that may interfere with the eye tracker's ability to locate the corneal reflection. However, eye tracking technology is improving; better image processing techniques, such as facial recognition, are helping to overcome these common calibration problems.

The general calibration procedure is composed of the following steps, to be performed by the operator after starting the system display/tracking program.

1. Move the application window to align it with the eye tracker's central calibration dot.
2. Adjust the eye tracker's pupil and corneal reflection threshold controls.
3. Calibrate the eye tracker.
4. Reset the eye tracker and run (program records the data).
5. Save recorded data.
6. Optionally calibrate again.

The final calibration step may be repeated to judge the amount of instrument slippage during the experimental trial (see Section 8.2 below). If the display/tracking program is one that relies on serial communication with the eye tracker (i.e., the program is responsible for generating the calibration stimulus points) the program must display the calibration stimulus at the same location the eye tracker is currently presenting to the viewer. This rather crucial application program requirement is discussed in the following section.

8.1 Software Implementation

If the application program is responsible for displaying the visual stimulus (i.e., the calibration dots and then later the test imagery) it is imperative that: (a) the program knows when to display the calibration dots or the test imagery, and (b) the calibration dots are aligned as precisely as possible so that the program's calibration dot is displayed at the location where the eye tracker expects it to be. The latter condition

---

[1] The eye tracker used to generate the images of Figure 8.1 is a binocular eye tracker mounted inside a Head-Mounted Display (HMD). Due to the proximity of the optics to each eye, two infra-red Light Emitting Diodes (LEDs) are used on either side of the eye. This is somewhat unusual; typically (for table-mounted, or remote eye trackers) one LED or other infra-red light source is used to project a single corneal reflection.

(a) Looking to upper-left.   (b) Looking to upper-right.

(c) Looking at screen center.

(d) Looking to bottom-left.   (e) Looking to bottom-right.

Fig. 8.1. Eye images during calibration (binocular eye tracking HMD).

is satisfied partly by appropriate mapping of coordinates between application program window (viewport) coordinates and the eye tracker screen coordinates, and the initial placement of the application program window on the display screen (e.g., the window is moved to the appropriate position on the display device such as the TV or within the HMD). The former condition is met by obtaining a status word from the eye tracker itself. It is therefore imperative that the eye tracker provide this valuable piece of data along with the POR value(s).

Assuming the eye tracker provides its current state (along with the current eye coordinates x and y), the pseudocode for the usual graphics drawing routine, modified to draw the calibration points at the appropriate time, is shown in Listing 8.1. This routine is sensitive to the three possible modes of the eye tracker: RUN, RESET, and CALIBRATE. A typical double-buffered display update routine, for systems without eye trackers, would normally just be expected to set up the scene projection and object transformation matrices, and then display the scene (stimulus) and swap buffers. In the modified routine, what is mostly needed is a way of drawing the calibration stimulus at the appropriate time.

```
sceneProjection();                              // e.g., ortho or perspective
sceneMatrix();                                  // model (object) transformation(s)
switch( eye tracker state ) {
   case RUN:
      if (displayStimulus)
         displayStimulus();                     // show image, VR, etc.
      break;
   case RESET:
   case CALIBRATE:
      drawCalibrationDot(x − 5, y − 5, x + 5, y + 5);
      break;
}
swapbuffers();                                  // swap buffers
```

Listing 8.1. Graphics draw/expose routine augmented with mode-sensitive eye tracking code.

Notice that in the pseudocode of Listing 8.1 the calibration stimulus is displayed in both RESET and CALIBRATE states. This facilitates the initial alignment of the application window with the eye tracker's initial positioning of its crosshairs. The default position is usually the center of the eye tracker screen. Provided the coordinate mapping routine is in place in the main loop of the program, the application program should correspondingly display its calibration dot at the center of its application window. The operator then simply positions the application window so that both points align in the eye tracker's scene monitor.

Notice also in Listing 8.1 that the stimulus scene (the image or VR environment) is only displayed if a display stimulus condition is satisfied. This condition may be set outside the drawing routine (e.g., in the main program loop) depending on some timing criterion. For example, if the experimental trial requires that an image be displayed for only five seconds, a timer in the main loop can be used to control the duration of the displayed stimulus by setting the display condition just after calibration has completed (state change from calibration to run). The timer then unsets the condition after the required duration has expired.

For VR applications, the draw routine may be preceded by viewport calls that determine which display, left or right, is drawn to. If the view is shifted horizontally, a binocular display is presented, otherwise, a biocular display is seen by the user. In VR, the calibration stimulus may require an orthographic projection call so that the 2D calibration points are not perturbed by a perspective transformation.

The main loop of the program is responsible for:

- Reading the data from the eye tracker (and the head tracker if one is being used)
- Mapping the eye tracker coordinates (and head tracker coordinates if one is being used)
- Starting/stopping timers if the experimental conditions call for a specific stimulus duration period
- Either storing or acting on the 2D (or 3D) gaze coordinates, if the application is diagnostic or gaze-contingent in nature, respectively

This general algorithm is shown as pseudocode in Listing 8.2. In this instance of the main loop, a timer of length DURATION is used to control the duration of stimulus display.

A similar loop is needed for a VR application. There are a few additional requirements dealing with the head position/orientation tracker and calculation of the three-dimensional gaze vector. The pseudocode for the main loop of a VR application is shown in Listing 8.3. The main loop routine for the VR application mainly differs in the calculation of the gaze vector, which is dependent on the stereoscopic geometry of the binocular system. The only other main difference is the data storage requirements. Instead of just the 2D point of regard, now the locations of both eyes (or the gaze vector) and/or the head need to be recorded. For gaze-contingent applications, instead of data storage, the gaze vector may be used directly to manipulate the scene or the objects within.

8.2 Ancillary Calibration Procedures

The calibration procedure described above is crucial for proper operation of the eye tracking device. Some devices will not even generate an eye movement data word until they have been calibrated. Device calibration is therefore necessary for any

```
while(true) {
  getEyeTrackerData(x, y);                              // read serial port
  mapEyeTrackerData(x, y);                              // map coordinates
  switch( eye tracker state ) {
    case RUN:
      if (! starting ) {
        starting  = true;
        startTimer();
        displayStimulus = true;
        redraw();                                       // redraw scene event
      }
      if ( checkTimer() > DURATION) {
        displayStimulus = false;
        redraw();                                       // redraw scene event
      } else {
        storeData(x,y);
      }
      break
    case RESET:
    case CALIBRATE:
      starting  = false;
      redraw();                                         // redraw scene event
      break;
  }
}
```

Listing 8.2. Main loop (2D imaging application).

subsequent eye tracker device operation. Additional calibration routines may be used to test the device accuracy slippage before and after experimental trials or perhaps for calibration of subsidiary software measures, possibly internal to the application program. Denoting the device calibration procedure as external, two ancillary procedures are presented, denoted internal, because these procedures are more related to the application software rather than to the eye tracker itself. The first such ancillary internal calibration procedure is described for checking the accuracy slippage of the eye tracker. This optional procedure is described for the 2D eye movement recording program described above. The second ancillary internal calibration procedure is described for 3D software calibration, as used in VR when binocular (vergence) eye movement parameters are not measurable a priori.

8.2.1 Internal 2D Calibration

Each experimental trial includes three calibration steps. Calibration is divided into two procedures: *external* and *internal*. External calibration pertains to the proprietary eye tracker instrument calibration procedure specified by the manufacturer. Internal

```
while(true) {
  getHeadTrackerData(eye,dir,upv);                    // read serial port
  getEyeTrackerData(x_l,y_l,x_r,y_r);                 // read serial port
  mapEyeTrackerData(x_l,y_l,x_r,y_r);                 // map coordinates

  // calculate linear gaze interpolant parameter s
  s = b/(x_l - x_r + b);

  // set head position
  h = [eye_x, eye_y, eye_z];

  // calculate central view vector
  v = [(x_l + x_r)/2 - x_h, (y_l + y_r)/2 - y_h, f - z_h];

  // multiply v by FOB matrix
  transformVectorToHeadReferenceFrame(v);

  // calculate gaze point
  g = h + sv;
  switch( eye tracker state ) {
    case RUN:
      if (! starting ) {
        starting  = true;
        startTimer();
        displayStimulus = true;
        redraw();                                     // redraw scene event
      }
      if (checkTimer() > DURATION) {
        displayStimulus = false;
        redraw();                                     // redraw scene event
      } else {
        storeData(x_l,y_l,x_r,y_r);
      }
      break
    case RESET:
    case CALIBRATE:
      starting  = false;
      redraw();                                       // redraw scene event
      break;
  }
}
```

Listing 8.3. Main loop (3D virtual reality application).

calibration pertains to the procedure developed within the developed application system for measurement of eye tracker accuracy. The eye tracker is externally calibrated using the vendor's proprietary 5- or 9-point calibration procedure. Internal calibration can be developed over any number of points, e.g., 30. The layout of the internal calibration points, denoted by the symbol + is shown in Figure 8.2. The point arrays are framed for clarity; horizontal and vertical lines do not appear during calibration. Figure 8.2 also shows the position of external calibration points overlaid on top of internal calibration points (in this case 9 points depicted by circles). Internal calibration is performed twice, immediately after external calibration (before stimulus display), and immediately after the stimulus display. Hence internal calibration is used to check the accuracy of the eye tracker before and after the stimulus display, to check for instrument slippage. The 30 internal calibration points can be shown in random order in a semi-interactive manner similar to the external calibration procedure. As each point is drawn on the screen (the subject is presented with a suitable simple stimulus, e.g., an $\times$ to minimize aliasing and flicker effects of an analog display), the application system waits for an input keypress before sampling eye movement data for a given sampling period (e.g., 800 ms). The input key delay allows the operator to observe eye stability on the eye tracker's eye monitor. The eye is judged to be stable once the eye tracker has repositioned the eye in the center of the camera frame. Recorded POR data is mapped to image coordinates in real-time. Calibration data can be stored in a flat text file for later evaluation.

Internal calibration procedures provide the basis for two statistical measures: the overall accuracy of the eye tracker, and the amount of instrument slippage during stimulus viewing. The latter measurement gives an indication of the instrument accuracy during the viewing task, i.e., by recording loss of accuracy between the before- and after-viewing calibration procedures. A simple experiment illustrates these measures. An average of recorded POR data points (centroid; following coordinate mapping) is obtained and the error between the centroid and calibration point is calculated. Each two-dimensional Euclidean distance measurement is converted to the full visual angle dependent on the viewing distance and calculated resolution of the television screen. A graphical example of this measurement is shown in Figure 8.3. The internal calibration locations are represented by +, sample measurements are represented by individual pixel dots, and centroid gaze positions are represented by circles, joined with the corresponding calibration point by a line. The length of the line is the average error deviation in pixels.

To quantify overall eye tracker accuracy succinctly, the average calibration error can be obtained from each set of calibration points in order to calculate an overall average statistic for the eye tracker. The resulting average instrument error is an average statistic over all calibration runs.

To quantify instrument slippage, statistical analysis can be performed on before- and after-viewing mean error measure. Figure 8.4 shows a composite plot of before and after internal calibration stimulus display. Notice that, in this example, measured

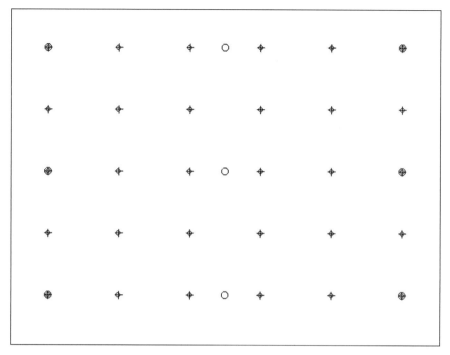

Fig. 8.2. Calibration stimulus: external (eye tracker) calibration points (circles) overlaid on internal calibration points (crosses).

eye positions generally correspond well to calibration points. To quantify this correspondence, a one-way ANOVA can be performed on the means of the before- and after-viewing average error measures. If no significant difference is reported between the means, one can consider instrument slippage to be nominal.

8.2.2 Internal 3D Calibration

For 3D gaze point estimation, determination of the scalar s (dependent on interpupillary distance, or baseline b) and focal distance f used in Equations (7.18)–(7.20) in Section 7.3 is difficult. Interpupillary distance cannot easily be measured in VR because the left and right eye tracking components function independently. That is, there is no common reference point. An internal 3D calibration procedure can be designed to estimate the interpupillary distance scaling factor s empirically (Duchowski et al., 2001).

The 3D calibration relies on a specially marked environment, containing nine clearly visible fixation targets, illustrated in Figure 8.5. Figure 8.5 shows scanpaths in a 3D virtual reality environment prior to eye movement analysis (see Chapter 12). Green

Fig. 8.3. Typical per-trial calibration data (one subject) after stimulus viewing (avg. error: 1.77°).

spheres in the figure represent raw GIP data. The nine × targets are distributed on five walls of the environment to allow head movement to be taken into account during analysis. Without a precise estimate of b and f, computed Gaze Intersection Points (GIPs) may appear stretched or compressed in the horizontal or vertical direction, as shown in Figure 8.5a. Following manual manipulation of b and f values, GIP data are recalculated and displayed interactively. The goal is to align the calculated GIP locations with the environmental targets on which the user was instructed to fixate during calibration. An example of this type of adjustment is shown in Figure 8.5b. Notice that the GIPs (represented by green transparent spheres) are now better aligned over the red targets than the raw data in Figure 8.5a. Once determined, appropriate scale factors are used to adjust each participant's eye movement data in all subsequent trials.

8.3 Summary and Further Reading

In summary, calibration of the eye tracker is essential toward proper eye movement data collection and analysis. The calibration procedure typically involves proper

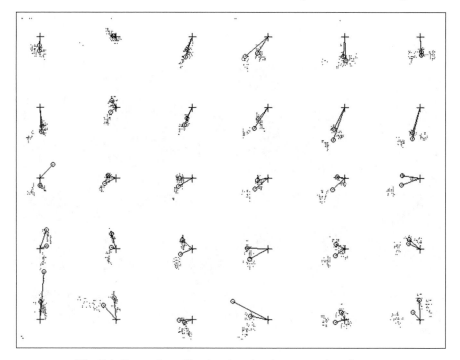

Fig. 8.4. Composite calibration data showing eye tracker slippage.

setting of the eye tracker's imaging optics (e.g., eye camera focus and contrast) and software threshold levels (e.g., pupil and corneal reflection). Although today's eye trackers have improved imaging and image processing components, traditional sources of calibration errors still exist and continue to pose problems. For example, contact lenses or eyeglasses may still interfere with some eye tracking devices; long eyelashes, heavy eye makeup, or "droopy" eyelids may prevent proper imaging of the eye, and in some scenarios the subject's own body and head movement may also cause imaging problems. In some instances, head stabilization (e.g., the use of a chin rest) may be required.

Unfortunately, there is no widely accepted "how-to" text explaining proper calibration procedures. Most often, the eye tracker manufacturer's usage manual is the best primary source of information. Furthermore, by far the best advice for good manual calibration is experience and practice. Proper usage of the eye tracker, and in particular fast and "good" calibration, often comes with repeated use of the device. A current goal of eye tracking research is to devise a system for which calibration is not necessary; i.e., manufacturers are currently pursuing the development of an autocalibrating eye tracker. Although the current goal of doing away with

(a) Prior to adjustment.

(b) Following adjustment.

Fig. 8.5. Adjustment of left and right eye scale factors.

calibration altogether has not yet been achieved, there is probably no better substitute for proper manual calibration than hands-on experience. It is often the case that new eye tracker operators are uncomfortable with the various controls of the device and may also feel nervous with directing human subjects to participate in eye tracking studies. However, with practice, one can quickly learn to properly use and calibrate an eye tracker. Once a sufficient comfort level is reached, proper calibration can be performed within a matter of seconds.

9

Table-Mounted System Hardware Installation

This chapter focuses on installation of a fourth-generation table-mounted eye tracker (cf. the analog-video, or third-generation table-mounted eye tracker discussed in Chapter 6). Although the modern counterpart is based on similar underlying principles as previous generation technology (video-based, corneal-reflection eye tracking), it is considerably easier to transport, install, and use. The table-mounted eye

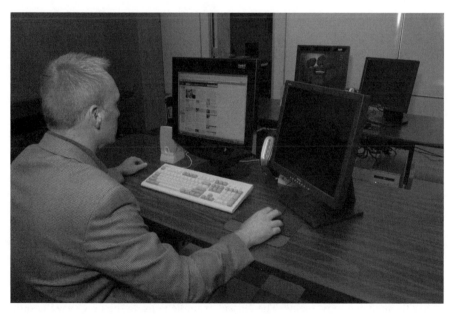

Fig. 9.1. Tobii dual-head eye tracking stations at Clemson University.

tracker may appear no different from a common flat panel display, and that is intentional. Unlike a typical monitor, however, a camera and infra-red LED optics are embedded beneath the LCD flat panel. The eye tracker on which this discussion is based is commercially available from Tobii Technology AB, based in Stockholm, Sweden. The dual-head installation at Clemson University, on which this and the remaining table-mounted system chapters are based, is shown in Figure 9.1. The particular hardware devices installed at Clemson are described here for reference.

Each of Clemson's eye tracking stations is centered on Tobii's ET-1750 eye tracker, which operates (samples) binocularly at 50 Hz at an accuracy of about $0.5°$ (bias error). The Tobii display is a 17 in. TFT flat panel running at 1280×1024 resolution. Tobii's eye tracking server runs on a dual 2.0 GHz AMD Opteron 246 Sun W2100z with 2 GB RAM running Windows XP. Its display is driven by an NVidia NVS280 graphics card. The application computer is a 2.2 GHz AMD Opteron 148 Ultra 20 with 1 GB RAM running Fedora Core 4 (Linux 2.6.11 with gcc v4.0.0). It is equipped with an NVidia GeForce 7800 GTX graphics card. Both computers are connected to a 1 Gb ethernet LAN. Both client and server machines are connected to a Belkin OmniView SOHO Series (F1DS104T) 4-Port KVM switch with audio PS/2 and USB support (firmware v2.0 7/7/2005). The KVM switch allows sharing of the keyboard, display, and mouse between the two computers.

9.1 Integration Issues and Requirements

The salient characteristic of the table-mounted eye tracker is its calculation and delivery of instantaneous (x, y) coordinates of the the user's gaze. Computing applications receiving this signal can then be developed to make use of this real-time information in a number of different ways (see Chapters 17–20). The most basic application is diagnostic in nature where the application collects gaze coordinates while the user is watching some form of stimulus (e.g., images, video, Web pages, desktop). Following eye movement (signal) analysis, one can infer the user's attentional strategies and in turn the stimuli's attentional qualities. Diagnostic applications are thus off-line because the display generally does not change contingent on the user's gaze location.

In contrast, interactive applications can make use of the viewer's real-time gaze location by either changing their appearance in some way or by directly invoking the user's gaze to affect control of the application. The former type of passive interface is termed *gaze-contingent* and is often used to evaluate characteristics of the human visual system. For example, one can provide imagery degraded in the user's peripheral visual field. If the user reports no perceived effect, then such a gaze-contingent display has successfully matched the human visual system's resolving capacity. The latter type of active interface allows the user to control aspects of the interface directly through eye movements. The classic example is using the (x, y) gaze coordinates in place of a manual mouse.

In general, most eye tracking devices allow the following functionality.

1. Connection: establish connection with the eye tracker (e.g., serial port or TCP/IP).

2. Calibration: display calibration points at the appropriate location and time.
3. Synchronization: display stimulus at the appropriate time (the eye tracker should be able to inform the application program of its state, or vice versa).
4. Data streaming: use eye tracker to capture data and/or update the stimulus scene in a gaze-contingent manner.

Two of the four key integration concerns identified in Chapter 6 have been for the most part removed, namely:

1. The capability of the eye tracker to provide fine-grained cursor control
2. The capability of the eye tracker to transmit its operating mode along with gaze (x, y) coordinates

Elimination of these two concerns is one of several reasons for the eye tracker's usability improvement. The fine-grained cursor control was previously used to obtain a mapping of the gaze coordinates over the display. This is still important, however, the Tobii transmits normalized gaze coordinates, i.e., in the range $[0, 1]$. Thus, mapping from the eye tracker's reference frame to a given application's is performed by simply scaling the normalized gaze coordinates by the extents of the application window. This greatly simplifies manipulation of gaze coordinates by the client application.

The second point concerning the tracker's operating mode has also been removed mainly due to the transfer of mode control to the application. That is, although now the host/client roles have been reversed, i.e., the Tobii is referred to as the server and the (interactive) application is now considered the client, it is the client application that controls what mode the eye tracker is in, i.e., idle, calibrating, or running. This transfer of control makes development of interactive applications considerably easier than with previous equipment. With older technology, a developer often needed a (very) patient secondary individual to serve as viewing subject when testing an eye tracking application. It was very difficult to self-calibrate because one had to control the eye tracker from a dedicated console, while operating the application from another. Two keyboards, mice, etc. were needed. Transferring control to the software allows the client application to control the eye tracker from within. Thus, self-calibration is now easy to perform.

The two other major integration concerns still exist, but due to Tobii's selection of a particular standardized display technology and release of Software Development Kit (SDK), the concerns have been greatly alleviated. They are:

1. Knowledge of the video format the eye tracker requires as input (e.g., currently VGA, hence no longer an issue)
2. Knowledge of the data format the eye tracker generates as its output (provided by the SDK reference)

Note that these two concerns, and particularly knowledge of the gaze data format and other details contained within the SDK, are mainly directed toward development of gaze-aware applications. This means that these issues are more of a concern for developers of interactive applications, or for those who wish to reinstrument their applications for off-line eye movement diagnostic analysis. For users who simply wish to evaluate currently available desktop applications (currently restricted to the Windows environment), images, Web pages, or videos, Tobii provides an even simpler solution with their ClearView software. ClearView is simply a Windows-based client application that Tobii has written that can collect eye movements over various forms of stimuli including Web pages (using Tobii's reinstrumented browser) and the Windows desktop (sacrificing a portion of the display refresh rate while obtaining data in this "screen" mode).

For diagnostic users of eye tracking technology, Tobii (as well as other manufacturers) have made great strides toward "plug-and-play" usability. Following basic installation, or rather simply computer-display connection (described below) and installation of eye tracking software, the experimenter is ready to begin collecting eye movements. The remaining details of Tobii's data format, SDK, etc. generally concern those wishing to develop their own gaze-aware applications (e.g., gaze-contingent displays).

9.2 System Installation

The Tobii ET-1750 eye tracker can be configured in several ways, one of which is acting as a server for a possibly remote eye tracking client application. Connecting over TCP/IP, a client application controls user calibration and then synchronizes with the eye tracker through Application Program Interface (API) callbacks. In the simplest case, both the client and server reside on the same machine; e.g., this would be typical of a single-PC such as a laptop running Windows, connected to the eye tracking display and camera and infra-red light optics.

An example configuration more suitable for development on a secondary application computer (e.g., Linux or Mac workstation or laptop) connected to the Tobii eye tracker is shown in Figure 9.2. Figure 9.2b shows the eye tracker display in isolation. Figure 9.2a shows the behind-the-scenes connections of the eye tracking station. An application computer drives the display's LCD flat panel. This computer is connected (via Ethernet) to the eye tracker computer processing images obtained from the camera embedded in the display. The eye tracking server also connects to the eye tracker, but not to its display, only its camera (e.g., via the IEEE 1394 or "firewire" interface) and infra-red lights (e.g., via the USB serial interface).

Note that the configuration in Figure 9.2 is a triple-head eye tracking station (in comparison, Figure 9.1 only shows a two-headed eye tracking station). Using all three monitors provides the simultaneous advantage of a wide-screen dual-head application (e.g., Linux) display and a monitor for the eye tracking server. The eye tracking server display may be somewhat extraneous, particularly if the server runs

(a) Schematic configuration.

(b) The ET-1750 flat panel.

(c) Typical use.

Fig. 9.2. Single Tobii eye tracking station hardware setup. From Ashmore et al. (2005) © 2005 Canadian Information Processing Society. Reprinted by permission.

as a background process and does not offer much in the way of a status display. Indeed, the lack of server status is expected because such status information should be obtained and displayed by the eye tracking client application. This is typical of the traditional client/server programming paradigm. For the triple-headed installation, both the eye tracking server and application client computers should be equipped with dual-headed graphics cards if one wishes to also perform diagnostic eye tracking work on the server machine.

Although the triple-headed installation may be somewhat of an overkill, eventually one must interact with the operating system on which the eye tracking server runs, and therefore, one must be able to connect the display to the eye tracking server. If a small monitor is not available, then the alternative is to share the eye tracking display between both server and client computers via a KVM, or Keyboard-Video-Mouse switch. The KVM switch is omitted from the schematic in Figure 9.2, as are the keyboard and mouse. The KVM switch is visible in Figure 9.1, however.

9.3 Lessons Learned from the Installation at Clemson

Compared to the previous eye tracking installation (see Chapter 6), setting up the current workstations was almost trivial, if not pleasurable. However, just as the previous installation had one troublesome component (video cable pinout differing between manufacturers), this installation was also not entirely trouble-free. Note that the video signal is still a concern, particularly if working with different manufacturers, e.g., Apple, who used to use specialized Apple Display Connectors. Contemporary video cards should support VGA, perhaps requiring a DVI-VGA adaptor (usually included with the display). If Tobii upgrades the display to DVI, the adapter would not be needed. Speculations aside, the video signal was a minor issue.

The component that required an appreciable amount of cajoling in the modern installation was the KVM switch. The specific KVM switch used shipped with a bug causing erroneous recognition of the mouse input when switching between Linux and Microsoft Windows. The symptom was a frozen mouse cursor requiring a reboot to unfreeze. The firmware update made available by the KVM switch manufacturer eventually alleviated the problem, but firmware installation was itself most annoying. The problem here was extremely poor feedback provided by the firmware upgrade utility. No status was given as the firmware was loaded and upon completion only an error was given. The error was apparently commonplace (as reported by technical support) and was meant to be ignored. However, to install the firmware, one had to disconnect all other signals to switch and connect only the computer loading the firmware. What's worse, the loading computer had to have a parallel port, which is something of a rarity. Thus, an older laptop had to be found which could install the firmware upgrade after everything else was disconnected.

Apart from the KVM switch annoyance, other pointers worth remembering are the typical requirements of sufficient power, network, keyboard, mouse, etc. When installing multiple multiheaded stations such as at Clemson (three stations are available for students), six network jacks were required. Depending on available IT support, this may or may not be a problem (at the Clemson installation, these jacks were not available initially, and a network switch was required).

9.4 Summary and Further Reading

This chapter presented key points for installation of a modern, table-mounted eye tracker. As with earlier head-mounted systems, successful installation still depends on appropriate signal routing and synchronization. Modern systems have simplified these concerns, however, particularly if signal acquisition is based on a networked client/server model. Rather than relying on the interpretation of vendor-specific serial data, communication over TCP/IP facilitates faster development due to the reliance of accepted standards, e.g., socket programming. Although the data received may still be specific to a particular vendor's format and protocol, their receipt from a socket stream is much simpler than byte assembly of a serial interface. An added benefit is the lifting of the proximity constraint placed by the length of a serial cable.

Video signal routing is still a (minor) concern, although this too has been simplified by the adoption of standardized displays, e.g., VGA. The installation of a three-headed system, described above, made simple by the use of a KVM switch, is a relatively complex one. Typical eye tracking "stations" need not be this elaborate and can for the most part resemble a typical computer workstation: computer and monitor along with eye tracking optics (possibly embedded).

Apart from hardware issues, the use of normalized screen coordinates (e.g., $(x, y) \in [0, 1]$) removes the problematic need for reference frame mapping, typically done in software, or at least the more daunting portion of coordinate registration that was needed previously. Given normalized point of regard coordinates, these are simply scaled to a given display or window over which eye movements are collected.

As with previous systems, the two primary sources where further information can be obtained on system installation and setup are the manufacturer's manual and any user community groups that may have assembled. Users of eye trackers still report descriptions of their apparatus and any specific technical innovations required for system development, specifically to foster replication of their experiments by other researchers. Reports can still be found in journal articles such as *Vision Research*, *Behavior Research Methods, Instruments, and Computers (BRMIC)*, and conference proceedings. There are various conferences that deal with eye tracking, either directly or indirectly. For example, conferences that deal with computer graphics (e.g., SIG-GRAPH, EuroGraphics, or Graphics Interface), human–computer interaction (e.g., SIGCHI), or virtual reality (e.g., VRST), still carry papers that discuss the use of eye trackers. Two main conferences dealing directly with eye tracking run on an inter-leaved biennial basis: the European Conference on Eye Movements (ECEM), and the U.S.-based Eye Tracking Research & Applications (ETRA). Finally, the eye move-ment email listservs *eye-movements* and *eyemov-l* are excellent on-line "gathering places" of eye tracker researchers.

10

Table-Mounted System Software Development

Virtually all the requirements for (interactive) eye tracking application development are similar today as they were a few years ago with older equipment (see Chapter 7), save for the most important: mapping of the eye tracker coordinates. If gaze coordinates provided by the tracker are normalized (as they are provided by the Tobii eye tracker), then mapping of the coordinates to the display is trivially accomplished by multiplying (scaling) the normalized coordinates by the screen extents. If an application is running in a window smaller than the screen, then one must obtain the window's position relative to the screen origin; this is usually readily available in most GUI toolkits. Given screen and window coordinates, mapping gaze coordinates to a given application's reference frame follows similar logic as in Chapter 7, namely based on linear interpolation.

Beyond coordinate reference frame alignment, the most important aspect of modern eye tracker programming (at least with the Tobii tracker) is that the application controls eye tracker functionality; it does not merely respond to changes in operation mode (where the modes are idle, calibrating, running). Furthermore, it is important to note that the eye tracking Application Program Interface (API) client library may need to run in its own concurrent thread. In the specific case of the Linux Tobii API, all API calls must be contained within the same thread. This is particularly relevant to traditional event-driven (GUI) programming because the Linux Tobii client library (discussed in detail below) is itself event-driven. This means that there are two simultaneous infinite loops running outside the programmer's control that must be synchronized in some way (in the Linux and Mac OS X environments this is accomplished with POSIX threads (Butenhof, 1997)). One loop is run by the GUI toolkit for which the programmer must implement functions (callbacks) to handle GUI events such as mouse motion, button presses, etc. The second loop is run by the Linux Tobii client thread for which the programmer must provide a callback function to handle its single event: consumption of gaze coordinates from the streaming buffer (this occurs when calibrating and running). The Linux Tobii API binary is available at <http://andrewd.ces.clemson.edu/tobii/>.

The Linux Tobii client API was ported to Linux at Clemson University in an independent effort (not written by Tobii Technology). Porting of the client code was achieved via TCP/IP socket programming (referencing the very readable texts of Comer and Stevens (2001, 1993)). The library was developed with Tobii's permission and was last tested with Tobii's TET Server v2.8.6 (ClearView Full Install Kit v2.5.1). The code may not function with future TET Server versions but will be ported to work with future TET Server versions if Tobii sanctions the effort and provides sufficient cooperation.

The remainder of this chapter discusses the Linux Tobii client API. Although it is specific to the Tobii eye tracker, it is expected that most eye trackers will contain similar functionality, for example connection to the eye tracker (TCP/IP or serial port), calibration, and data streaming.

10.1 Linux Tobii Client Application Program Interface

Table 10.1. Linux Tobii client API function listing.

Tet_Init	Tet_CalibClear	Tet_SynchronizeTime
Tet_Connect	Tet_CalibLoadFromFile	Tet_PerformSystemCheck
Tet_Disconnect	Tet_CalibSaveToFile	Tet_GetSerialNumber
Tet_Start	Tet_CalibAddPoint	Tet_GetLastError
Tet_Stop	Tet_CalibRemovePoints	Tet_GetLastErrorAsText
	Tet_CalibGetResult	
	Tet_CalibCalculateAndSet	

The Linux Tobii client API consists of the C functions listed in Table 10.1. The functions in the left column of Table 10.1 are used to initialize the eye tracking client thread, set up the TCP/IP connection, and start and stop the eye tracking run (data streaming). The functions listed in the middle column of Table 10.1 all have to do with eye tracker calibration. Note that nothing gets drawn as a side effect; only calibration coordinates are passed to the eye tracker. Transmission of calibration coordinates must synchronize with the drawing thread so that calibration stimulus targets (e.g., dots) are drawn at the same coordinates as those transmitted and at the same time. One way to do this is via POSIX threads' conditional waits, as explained in Chapter 11. The functions listed in the right column of Table 10.1 are auxiliary support functions dealing with real-time synchronization, error logging, and other information gathering.

Note that Tet_Start and Tet_CalibAddPoint require the specification of an event handling (callback) function (see below for details) and thus transfer control to that event handler. Stoppage of an eye tracking run, performed via Tet_Stop, can only be performed from within that callback event handler (or via the setting of a flag to issue Tet_Stop from another thread). These as well as the remaining "Tet" function calls are explained in detail below.

10.1.1 Tet_Init

```
long Tet_Init ( void );
```

Listing 10.1. Tet_Init declaration.

The Tet_Init call must be called once at thread initialization. This sets up thread-specific data and initializes the internal timer library. (This function is only relevant in the Linux/Mac OS X environments.)

Example: Tet_Init();

10.1.2 Tet_Connect, Tet_Disconnect

```
long Tet_Connect(  char*          pServerAddress,
          unsigned short portnumber,
          char*       pDebugLogFile);
long Tet_Disconnect(void);
```

Listing 10.2. Tet_Connect and Tet_Disconnect declarations.

The Tet_Connect call connects to a (possibly remote) Tobii eye tracking server. The argument pServerAddress is a null-terminated array of characters that contains the eye tracking server's host's name or IP address. If, for example, the server were running on the local machine, pServerAddress would be set to either 127.0.0.1 or localhost. The argument portnumber should be set to 4455 (or NULL, which defaults to 4455). The argument pDebugLogFile is the pathname to a log file. Generally, Tet_Connect must be called prior to calling any other Tet_ function.

Tet_Disconnect disconnects from the eye tracking server. Each call should only be made once.

Example: Tet_Connect("tobii.cs.clemson.edu",4455,"logfile");
Example: Tet_Disconnect();

```
long Tet_Start( Tet_CallbackFunction  func,
        void*              pAppData,
        unsigned long        interval );
long Tet_Stop(void);
```

Listing 10.3. Tet_Start and Tet_Stop declarations.

10.1.3 Tet_Start, Tet_Stop

Tet_Start is a blocking function that starts gaze data streaming. When new data are available the callback function func is called immediately with fresh data (see Tet_CallbackFunction below). The argument pAppData is optional and points to a user-defined structure that can be passed to func. The interval argument denotes an interval in milliseconds when the callback function will be called; if it is set to 0, the callback will never be called for a timer reason (but will be called when new gaze data are available).

The callback function may be set up for periodic calls even though there are no new gaze data. Set the interval parameter to something other than 0 and the callback function will be called with **ETet_CallbackReason** set to TET_CALLBACK_TIMER and pData set to NULL.

Example: Tet_Start(gazeDataReceiver,NULL,0);
Example: Tet_Stop();

10.1.4 Tet_CalibClear, Tet_CalibLoadFromFile, Tet_CalibSaveToFile, Tet_CalibAddPoint, Tet_CalibRemovePoints, Tet_CalibGetResult, Tet_CalibCalculateAndSet

Function Tet_CalibClear is used when calibration is to be performed. This clears all points on which Tet_CalibCalculateAndSet bases its calculations. Alternatively, function Tet_CalibRemovePoints can be used to clear only those points within a radius from the point at coordinates (x,y). Units for (x,y) and radius range from 0 to 1 where (0,0) is the screen's upper left and (1,1) is the screen bottom right. The argument **ETet_Eye** is one of TET_EYE_LEFT, TET_EYE_RIGHT, or TET_EYE_BOTH.

The blocking function Tet_CalibAddPoint is used to perform the calibration. This function adds a new calibration point at coordinates (x,y). Normal usage is to simultaneously display something on the screen at the same (normalized) coordinates, making sure that the viewer gazes at this point at this time. Then call Tet_CalibAddPoint to make the eye tracker record calibration samples at this location. The eye tracker will obtain nrofdata calibration samples at this calibration point. With a larger nrofdata, the longer the calibration point will be displayed. If

```
long Tet_CalibClear(void);
long Tet_CalibLoadFromFile(char* pFile);
long Tet_CalibSaveToFile(char* pFile);
long Tet_CalibAddPoint( float          x,
            float          y,
            unsigned long      nrofdata,
            Tet_CallbackFunction  func,
            void*          pAppData,
            unsigned long      interval );
long Tet_CalibRemovePoints( ETet_Eye eye,
              float    x,
              float    y,
              float    radius);
long Tet_CalibGetResult(    char*         pFile,
            STet_CalibAnalyzeData* pData,
            long*          pLen);
long Tet_CalibCalculateAndSet(void);
```

Listing 10.4. Tet_CalibClear, Tet_CalibSaveToFile, Tet_CalibAddPoint, Tet_Calib-
RemovePoints, Tet_CalibGetResult, Tet_CalibCalculateAndSet declarations.

nrofdata is too large, the entire calibration may be too time consuming. Generally, five calibration points displayed at the screen center and screen corners are adequate (although as few as two calibration points at two opposing corners may be sufficient). The total number of calibration samples must not exceed 200. A good number of calibration samples per point seems to be in the 16–24 range. With five calibration points, this leads to a total of 80–120 calibration samples.

The function Tet_CalibAddPoint behaves in a manner similar to Tet_Start in that it blocks and calls the **Tet_CallbackFunction** func to consume gaze data.

When all calibration points have been displayed by the application program and sampled by the eye tracker, the function Tet_CalibCalculateAndSet must be called to invoke the new calibration.

Function Tet_CalibGetResult retrieves diagnostic information regarding a particular calibration. If pFile is NULL, results from the current calibration will be returned. Data returned include the target calibration points, the resulting sampled mapped points, and an indication if the calibration point was discarded for some reason (see Listing 10.5 for specification of the **STet_CalibAnalyzeData** data structure). The argument pLen will be altered by this function to indicate the number of elements returned in pData. If the returned data exceed the initial dimension of pData, pData will be filled to its limit, making it possible to reallocate pData to its true size and reissuing the call to Tet_CalibGetResult. If pFile is not null but a null terminated array of characters containing the path and filename to a personal calibration file

```
typedef struct _STet_CalibAnalyzeData {
    float       truePointX;
    float       truePointY;
    float       leftMapX;
    float       leftMapY;
    long        leftValidity  ;
    float       leftQuality ;
    float       rightMapX;
    float       rightMapY;
    long        rightValidity  ;
    float       rightQuality ;
} STet_CalibAnalyzeData;
```

Listing 10.5. STet_CalibAnalyzeData data structure.

saved by Tet_CalibSaveToFile, diagnostic (error) information from that calibration will be returned.

The function Tet_CalibSaveToFile provides the capability to save a calibration run to a file. This way, a returning user may read in the previous calibration via Tet_CalibLoadFromFile without having to recalibrate (provided the seating position is roughly the same).

Example: Tet_CalibClear(void);
Example: Tet_CalibLoadFromFile(myCalibration);
Example: Tet_CalibSaveToFile(myCalibration);
Example: Tet_CalibAddPoint(x,y,samples,gazeDataReceiver,NULL,0);
Example: Tet_CalibRemovePoints(TET_EYE_BOTH,x,y,radius);
Example: Tet_CalibGetResult(NULL,pData,255);
Example: Tet_CalibCalculateAndSet();

10.1.5 Tet_SynchronizeTime, Tet_PerformSystemCheck

```
long Tet_SynchronizeTime( unsigned long*      pSec,
                 unsigned long*      pMicroSec);
long Tet_PerformSystemCheck(void);
```

Listing 10.6. Tet_SynchronizeTime and Tet_PerformSystemCheck declarations.

The function Tet_PerformSystemCheck is used to obtain the current status of the Tobii eye tracking server; it is mainly used to find out if the camera and diodes are attached.

The function Tet_SynchronizeTime sets up the Tobii server's time module and should enable local routines to convert local timestamps to and from the Tobii server's timestamps. Arguments pSec and pMicroSec should denote the difference between the local (client-side) time and the remote (server-side) time.

Note: Depending on how the client's and server's hardware clocks differ, drift causing an error is likely to increase as time goes. If the Tobii server is running on the same host (client-side), this will be detected and handled properly only if the loopback interface was used (IP address 127.0.0.1) to connect. If any other address is used to connect to localhost, an unnecessary error will be introduced in the timestamps.

In the Linux environment, as of version v2.8.6, these functions were not fully tested. It seemed, however, that the time obtained via Tet_SynchronizeTime should at least be useful in establishing the network lag, assuming the data returned denotes the difference between local and remote hosts. However, this was never fully confirmed and so should be used with caution.

Example: Tet_SynchronizeTime(&pSec,&pMicroSec);
Example: Tet_PerformSystemCheck();

10.1.6 Tet_GetSerialNumber, Tet_GetLastError, Tet_GetLastErrorAsText

```
long Tet_GetSerialNumber( char* pSerialDiodeController,
                char* pSerialCamera);
long Tet_GetLastError(void);
long Tet_GetLastErrorAsText(char* pError);
```

Listing 10.7. Tet_GetSerialNumber, Tet_GetLastError, and Tet_GetLastErrorAsText declarations.

Function Tet_GetSerialNumber obtains the hardware serial numbers, if present, from the diode controller and the camera. Arguments pSerialDiodeController and pSerialCamera are null terminated arrays of characters that will never be larger than 64 bytes including the null terminator. Both arrays must be allocated by the caller.

Whenever a TET_ERROR is returned from any Tet_ function, a call to Tet_GetLastError will return the specific last error. The error will be overwritten as soon as another error occurs.

Note: The Tet_GetLastError function was never fully tested in the Linux environment; function Tet_GetLastErrorAsText should be used instead.

Function Tet_GetLastErrorAsText behaves similarly to function Tet_GetLastError except the error is copied to argument pError, a null terminated array of characters that will never require more than 255 bytes.

Example: Tet_GetSerialNumber(SerialDiodeController,SerialCamera);
Example: Tet_GetLastError();
Example: Tet_GetLastErrorAsText(Error);

10.1.7 Tet_CallbackFunction

```
typedef void (*Tet_CallbackFunction)( ETet_CallbackReason reason,
                      void*        pData,
                      void*        pAppData);
```

Listing 10.8. Tet_CallbackFunction declaration.

The ***Tet_CallbackFunction**'s **ETet_CallbackReason** argument will be set to TET_CALLBACK_GAZE_DATA or TET_CALLBACK_TIMER. The optional argument pAppData can point to any user-defined structure, and the argument pData should be cast to **(STet_GazeData *)** if the reason for the callback was set to TET_CALLBACK_GAZE_DATA (see Listing 10.9 for specification of the **STet_GazeData** data structure).

10.2 A Simple OpenGL/GLUT GUI Example

In this section a simple OpenGL/GLUT example is presented, focusing on thread creation and basic display. Because calibration synchronization is slightly more complicated, it is described in greater detail in the next chapter.

The basic steps of a simple Linux eye tracking application involve:

1. Initializing the GUI, in this case GLUT; setting up the GUI and eye tracking (ET) threads
2. Specifying the main ET thread and GUI drawing loop

In this example, the application simply draws the real-time gaze point on the blank screen as the user looks at the screen. Three gaze points are rendered: the left and right gaze points (represented by a green and red dot, respectively, although this is not detailed in the given listings), and the average gaze point represented by a blue

```
typedef struct _STet_GazeData {
    unsigned long  timestamp_sec;
    unsigned long  timestamp_microsec;
    float          x_gazepos_lefteye;
    float          y_gazepos_lefteye;
    float          x_camerapos_lefteye;
    float          y_camerapos_lefteye;
    float          diameter_pupil_lefteye ;
    float          distance_lefteye ;
    unsigned long  validity_lefteye   ;
    float          x_gazepos_righteye;
    float          y_gazepos_righteye;
    float          x_camerapos_righteye;
    float          y_camerapos_righteye;
    float          diameter_pupil_righteye;
    float          distance_righteye;
    unsigned long  validity_righteye  ;
} STet_GazeData;
```

Listing 10.9. STet_GazeData data structure.

dot. Also omitted from the listings are two additional feedback disks that are drawn to indicate the user's head position relative to the display. For brevity, OpenGL calls are omitted and simply replaced by draw_point(x,y). In the actual example program, simple points are drawn.

Listing 10.10 shows the main routine where the GLUT GUI toolkit is initialized, the Tobii thread is created, and flow of control is given up to the GUI after specifying an idle loop main_loop that executes whenever there is no pending GUI event. The object state serves as a repository of global variables, similar to a C++ singleton pattern.

Listing 10.11 sketches the main features of the GUI's idle loop. There are two main drawing states for when the eye tracker is calibrating or running. When calibrating, calibration dots are drawn in a synchronized fashion with the ET thread (see Chapter 11 for details). When running, simple OpenGL points are drawn at the coordinates of the left, right, and average eye gaze location. Not shown in Listing 10.11 are two translucent disks (gluDisks) that are drawn to indicate the (mirrored) position of the eyes with respect to the camera, although this type of visual feedback is very useful for proper self-positioning by the user.

Listing 10.12 is the top-level sketch of the ET thread. It is a POSIX thread, and is initialized to allow asynchronous cancellation (by the parent thread). The first task is to initialize internal Tobii state variables via Tet_Init. The remainder of the thread is a loop that contains the primary eye tracking modes of calibrating and running. Note

```
int main(int argc, char **argv)
{
    pthread_t   thread;

    // object that holds global state (like singleton pattern)
    state = new State();

    // initialize GLUT, create window
    glutInit (&argc, argv);
    glutInitDisplayMode(GLUT_DOUBLE|GLUT_RGBA|GLUT_DEPTH);
    glutCreateWindow("Demo");

    // setup GUI callbacks and main_loop
    glutDisplayFunc(on_expose);
    glutReshapeFunc(on_resize);
    glutKeyboardFunc(on_key_press);
    glutMouseFunc(on_button_press);
    glutMotionFunc(on_motion_notify);
    glutIdleFunc(main_loop);

    // initialize mutex for access to global state info
    pthread_mutex_init(&state→mutex,NULL);

    // create tobii thread (this sets tobii thread running)
    pthread_create(&thread,NULL,&tobii_thread,NULL);

    // glut main loop
    glutMainLoop();

    // thread and mutex cleanup
    pthread_cancel(thread);
    pthread_mutex_destroy(&state→mutex);

    return(0);
}
```

Listing 10.10. GUI initialization and Tobii thread creation.

that in this GLUT implementation, the user controls transitions into these modes by selecting menu items that change the state→status bitfield and thereby mode transitions.

Listing 10.13 shows a typical connection to the eye tracking host (server). Note that because the state object is shared between the GUI and ET threads, any alteration (writing) of its internal variables must be treated as a critical section. Thus, changing its status bitfield is performed only after locking the POSIX mutex (that is itself

```
void main_loop(void)
{
  // clear screen
  ...

  // view transformation
  ...

  // object transformation
  ...

  if (state→status.connected) {
    if (state→status.calibrating &&
        state→status.calibstarted) {

      // block until Tobii thread advances to new point

      // draw calibration dot at (x, y) coordinates

      // signal Tobii thread to advance to new point

    }
    else if (state→status.running) {
      // draw left gaze point (normalized coordinates scaled to window size)
      draw_point(state→xₗ * w, h−state→yₗ * h);

      // draw right gaze point
      draw_point(state→xᵣ * w, h−state→yᵣ * h);

      // draw average gaze point
      x = (state→xₗ+state→xᵣ)/2 * w;
      y = (state→yₗ+state→yᵣ)/2 * h;
      draw_point(x, h − y);

      // draw current camera eye point (left eye, right eye, average of both)
      ...
    }
  }
  glutSwapBuffers();
}
```

Listing 10.11. GUI main loop.

```
void* tobii_thread (void* ap)
{
   // allow thread to be canceled (asynchronously, e.g., immediately)
   pthread_setcancelstate(PTHREAD_CANCEL_ENABLE,NULL);
   pthread_setcanceltype(PTHREAD_CANCEL_ASYNCHRONOUS,NULL);

   // initialize Tobii (this is critical — sets up thread-specific data)
   Tet_Init ();

   while(!state→status.finished) {
      if (! state→status.connected)
         if (state→status.connect) tobii_connect();
      else {
         if (state→status.calibrating) tobii_calibrate  ();
         else if (state→status.running) tobii_run();
         else if (state→status.disconnect) tobii_disconnect();
      }
   }

   // clean exit
   pthread_exit(NULL);
}
```

Listing 10.12. Tobii thread.

stored in the state object).

```
void tobii_connect()
{
   Tet_Connect(state→tobiiHost,state→tobiiPort,"logfile");

   pthread_mutex_lock(&state→mutex);
   state→status.connected = 1;
   state→status.connect = 0;
   pthread_mutex_unlock(&state→mutex);
}
```

Listing 10.13. Tobii thread: connect to eye tracker.

Listing 10.14 shows the calibration pseudocode. Tet_CalibClear is called once per calibration to clear out all calibration information on the eye tracking server. Once this is accomplished, successive calibration point coordinates are sent to the eye tracking server one after another. Synchronization with the GUI thread is discussed in detail in Chapter 11. Once all point coordinates have been sent, Tet_CalibCalculateAndSet is called to complete the calibration. At this point, Tet_CalibGetResult could be called to examine internal calibration error.

Listings 10.15 and 10.16 show the use of the Tet_Start and Tet_Disconnect calls, respectively. Note that Tet_Start is a blocking function and transfers control to the eye tracking client callback, in this case the gazeDataReceiver function, shown in Listing 10.17.

Listing 10.17 consumes gaze data as they become available and copies the data to the shared state object. If it is detected that the operating mode has transitioned to stop (idle) mode, a call to Tet_Stop is made, returning control to the point where Tet_Start was called.

10.3 Caveats

It is important to re-emphasize the concurrency between the Tobii client thread and the rest of the program, e.g., GUI and/or display threads. The eye tracking client thread is event-driven; the server communicates to the client thread via the **Tet_CallbackFunction** callback (i.e., eye movement coordinates are consumed by the client thread's callback routine). This effectively means that during eye movement data collection (or calibration!) the client thread is blocked (because control is effectively limited to one function call, the **Tet_CallbackFunction** routine). If other programming tasks are needed (e.g., screen display redraws, or GUI event handling), this functionality must be placed in a separate thread. Note that GUI toolkits are also event-driven and therefore program control flow is given up to the GUI process. Synchronization between processes is therefore crucial.

The above caveat refers to the Linux environment. Tobii (and other manufacturers) provide Windows-based Software Development Kits, or SDKs, that may function completely differently. That is, concurrency and synchronization may still be present, but may be transparent to the programmer.

10.4 Summary and Further Reading

This chapter described one method for development of an interactive eye tracking application. The discussion centered on a specific eye tracker model (Tobii's ET-1750) in the Linux environment. It is likely that other eye tracking manufacturers have similar Application Program Interfaces (APIs) to their trackers. It is expected that most eye trackers, which essentially are just motion trackers (e.g., similar in a sense to 3D head motion trackers), will provide analogous programming methods for connection, calibration, and data streaming. The manufacturers' specifications for their hardware and software are therefore the best source of information. If one is

```
tobii_calibrate ()
{
  if (! state→status.calibstarted) {

    // clear calibration — do this only once per calibration!
    Tet_CalibClear();

    pthread_mutex_lock(&state→mutex);
    state→status.calibstarted = 1;
    pthread_mutex_unlock(&state→mutex);
  }

  if ( state→status.calibstarted) {

    // signal gui to draw calib point

    // get eye tracker to start recording at this point
    Tet_CalibAddPoint(x,y,samples,gazeDataReceiver,NULL,0);

    // block for gui to finish drawing

    // advance to next calibration point
    pthread_mutex_lock(&state→mutex);
    state→calpoint++;
    pthread_mutex_unlock(&state→mutex);

    if ( last calibration point ) {

      // signal gui to exit calibration

      pthread_mutex_lock(&state→mutex);
      state→status.calibstarted = 0;
      state→status.calibrating = 0;
      state→calpoint = 0;
      pthread_mutex_unlock(&state→mutex);

      // finished calibrating, tell the eye tracker we're done
      Tet_CalibCalculateAndSet();

      // hint: should perform a quality check here and save calibration
      // file for future use if calibration is good enough
    }
  }
}
```

Listing 10.14. Tobii thread: calibrate.

```
void tobii_run ()
{
  if (! state→status.runstarted) {
    // start the subscription of gaze data — Tet_Start will not return
    // until Tet_Stop is called or there is an error, therefore, the only
    // place to to stop the eye tracker is in the gazeDataReceiver callback.
    pthread_mutex_lock(&state→mutex);
    state→status.runstarted = 1;
    pthread_mutex_unlock(&state→mutex);

    Tet_Start (gazeDataReceiver,NULL,0);
  }
}
```

Listing 10.15. Tobii thread: run.

```
void tobii_disconnect ()
{
  Tet_Disconnect();

  pthread_mutex_lock(&state→mutex);
  state→status.connected = 0;
  state→status.disconnect = 0;
  pthread_mutex_unlock(&state→mutex);
}
```

Listing 10.16. Tobii thread: disconnect from eye tracker.

actively researching different kinds of eye trackers, it may be worthwhile to inquire about the tracker's programming accessibility prior to purchase. Indeed, the Linux programming environment for the Tobii discussed above was not developed by Tobii. Instead, Tobii emphasizes and provides reference materials for their Windows-based Software Development Kit (SDK). This SDK offers Windows client code, functionally similar to the above Linux library, in C# and and C++.

For further development of interactive or diagnostic applications under Linux or Mac OS X, refer to the actual code examples available online: <http://andrewd.-ces.clemson.edu/tobii/>. There, examples include the above simple **glut-pthreads** program, as well as simple diagnostic eye movement capture over still imagery (the **gtk** example, which is a port of the older ISCAN eye movement capture program), as well as interactive, gaze-contingent display (the **gcd-img++** example) and fisheye (**glut-pdt**) examples.

```
void gazeDataReceiver(ETet_CallbackReason reason,
                      void*                pData,
                      void*                pApplicationData)
{
    STet_GazeData* pGazeData=NULL;

  pGazeData = (STet_GazeData *)pData;

  if ( state→status.connected) {

    switch(reason) {
      case TET_CALLBACK_GAZE_DATA:
        if ( state→status.running) {

            // copy to state data structure
            pthread_mutex_lock(&state→mutex);
            state→xl = pGazeData→x_gazepos_lefteye;
            state→yl = pGazeData→y_gazepos_lefteye;
            state→xr = pGazeData→x_gazepos_righteye;
            state→yr = pGazeData→y_gazepos_righteye;
            state→cxl = pGazeData→x_camerapos_lefteye;
            state→cyl = pGazeData→y_camerapos_lefteye;
            state→cxr = pGazeData→x_camerapos_righteye;
            state→cyr = pGazeData→y_camerapos_righteye;
            pthread_mutex_unlock(&state→mutex);
        }
        break;
      case TET_CALLBACK_TIMER:
        break;
    }

    if ( state→status.running && state→status.stop) {

      Tet_Stop();

      pthread_mutex_lock(&state→mutex);
      state→status.running = 0;
      state→status.runstarted = 0;
      state→status.stop = 0;
      pthread_mutex_unlock(&state→mutex);
    }
  }
}
```

Listing 10.17. Tobii thread: gazeDataReceiver.

Table-Mounted System Calibration

From a user's point of view, calibration of modern eye trackers has improved considerably. The most noticeable improvement is the absence of a chin rest. By allowing constrained head movement (usually within a finite volume), modern eye trackers, such as the Tobii, have rendered the chin rest unnecessary (at the expense of some gaze accuracy, compared with, for example, dual-Purkinje eye trackers that require bite bars). However, even though calibration has been greatly simplified (a most welcome technological advancement to be sure), it is still susceptible to the same problems that plagued older technology (i.e., interference from eyelashes, certain spectacle rims, etc.). This is of course due simply to interference in the camera's image of the eye(s), just as electromagnetic head trackers suffer from interference from nearby metallic surfaces. Because both older and modern table-mounted trackers rely on video cameras to image the user's eye(s), it is still important that users sit at an appropriate distance and level from the camera optics. In some ways, the complexity of calibration has been shifted from the user to the application developer.

Ensuring that users sit at an appropriate position relative to the imaging optics is now mainly a matter of proper visual feedback/notification. For example, Tobii includes a track status window that displays the location of the user's eyes from the camera's point of view. Figure 11.1 shows three instances of real-time feedback: a user sitting nearly optimally in front of the camera at an appropriate level, sitting slightly too far (with the head dropping in elevation as a result) with the head tilted, and with one eye closed. Providing this type of feedback is the responsibility of the programmer although it is not difficult. The most difficult aspect may be when displaying such feedback is appropriate. Should it be provided continuously, even during diagnostic eye movement capture? This is clearly a design decision and depends on the nature of the application being developed. For applications where eye tracking feedback may be a distraction, it seems natural to restrict gaze point and camera eye position display to the calibration sequence, because in this case some form of visible targeting is required anyway.

Fig. 11.1. Tobii eye tracking status window: sitting level; too far back with head tilted; left eye closed.

Another tradeoff in calibration complexity appears in the form of eye tracker configuration (more or less). That is, a secondary human operator is no longer needed to manipulate camera controls such as focus, aperture, etc. (at least with trackers such as Tobii's). This is also a welcome advancement, however, calibration from a programmer's perspective is complicated slightly due to the concurrency of the eye tracking (ET) client and event-driven GUI paradigms. That is, because the application now drives the eye tracker and not vice versa (this is highly preferable to the contrary operation of older systems), the application must synchronize both ET and GUI program components. Thus, operational control is gained at the expense of programming complexity.

Synchronization of ET and GUI components can be fairly easily accomplished via mutual exclusion in the program's critical sections. This is a common approach in concurrent programming. Whenever concurrent program components (e.g., threads) write to shared memory, exclusive access can be guaranteed by the use of mutexes (or semaphores). The code listings in Chapter 10 made frequent use of the lone state→mutex to access the shared state→status bitfield. However, details of calibration synchronization were purposefully postponed to the following sections.

Note that calibration synchronization as described below may perhaps be somewhat overly complicated. Instead, the use of a few more additional bitfields may be sufficient. In this chapter, calibration synchronization is accomplished via conditional waits.

11.1 Software Implementation

The design for calibration synchronization with conditional variables was based on Butenhof's (1997) client/server programming example. In Butenhof's client/server system, a client requests that a server perform an operation asynchronously on a data element, while the client either waits for the server or proceeds in parallel and checks for the result of the operation at a later time. Note that it would be simplest to force the client to wait for the server by, for example, blocking on a mutex.

The intuitive blocking strategy blocks the ET thread while the GUI thread draws the calibration dot. With drawing completed and with the ET thread unblocked, the ET thread notifies the eye tracker to sample at the calibration location. The trouble with this simple approach is that calibration sampling occurs after the dot has been drawn. If the delay between drawing completion and calibration sampling is excessive, the user, out of boredom or anticipation for the next dot, may look away from the calibration location and thus spoil the calibration. The obvious criterion during calibration is ensuring that the user look at the calibration dot when the eye tracker is sampling at that same location. This may be accomplished by issuing a verbal instruction to this effect to the user prior to calibration. Or, as is suggested here, use calibration dot animation as a means of orienting the user's attention. That is, animate the dot and instruct the eye tracker to sample at the same time.

Adopting Butenhof's client/server model, with the ET thread taking on the role of the client and the GUI thread taking the role of server, the idea is for the ET thread to request the GUI thread to animate a calibration dot, while the ET thread signals the eye tracker to sample at the same calibration coordinates. Thus, ideally, the calibration dot is being drawn at the same time the eye tracker is calibrating.

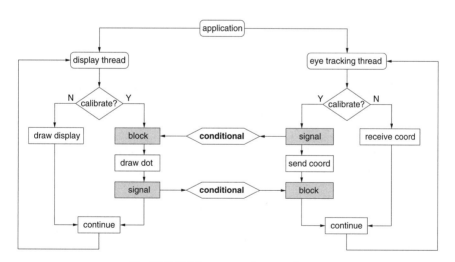

Fig. 11.2. Tobii concurrent process layout.

The concurrent approach to drawing and calibration sampling is depicted in Figure 11.2. The ET thread sends its request to the eye tracker to sample at the given calibration coordinates immediately after the ET thread signals the GUI thread to start drawing the calibration dot. The GUI thread animates the dot by, for example, pulsating the dot's size to attract and hold attention.

The concurrency between the ET and GUI threads should be considered carefully. It cannot be assumed that animating the calibration dot is either slower or faster than calibration sampling. It should be clear, however, that once each task is completed by its respective thread, that thread would continue with its next statement. Without a subsequent resynchronization following completion of both tasks, something of a race condition would then ensue. For example, if the GUI thread was faster in drawing the calibration dot than the eye tracker's calibration sampling, the GUI thread would continue to draw the next dot in the sequence. If the GUI thread was much faster than the ET thread, it is possible that it could complete drawing of all dots before the eye tracker had a chance to finish sampling at the very first coordinates. Conversely, if the ET thread was much faster, the eye tracker would be sampling at successive calibration coordinates although the user would still be looking at the very first calibration dot.

To resynchronize both threads after drawing/sampling of a particular calibration co-ordinate, the ET thread blocks after it has completed sampling at the calibration coordinates. Recall that the function call that is issued by the ET thread, namely Tet_CalibAddPoint, is blocking. Thus, we know the eye tracker has completed its calibration sampling when this statement completes. At this point the ET thread blocks itself and waits for the GUI thread to complete drawing the calibration dot. Listing 11.1 highlights the relevant signal and blocking calls issued by the ET thread. Only the highlighted synchronization code differs from Listing 10.14 given previously.

Consider again the GUI thread. In the preceding discussion, it is tacitly assumed that the GUI thread takes longer than the ET thread in calibrating at the given coordinates. What happens if the GUI thread completes drawing before the eye tracker completes sampling? After drawing, the GUI thread signals the ET thread to unblock and then carries on in the loop to the next calibration point. If the next coordinates in the sequence were available, the GUI thread would race ahead of the ET thread. To prevent this race condition, it is blocked before it can start drawing the next dot. In fact, Butenhof's client/server example was chosen because it not only provides a means of synchronization between the processes, it also allows communication of the calibration coordinates. The ET thread controls which calibration point is to be sampled next and passes that information to the GUI thread. If this information is unavailable, the GUI thread must block and wait. Listing 11.2 highlights the relevant blocking and signaling calls made by the GUI thread. The highlighted synchronization code is the only difference from Listing 10.11 given previously.

The synchronization code in Listings 11.1 and 11.2 uses a daimon object to control the interprocess calibration synchronization and communication. The *daimon* moniker was chosen in place of Butenhof's *server* to avoid confusion between the eye tracking server and to differentiate from the traditional (UNIX) system *daemon*.

```
tobii_calibrate  ()
{
  if (! state→status.calibstarted) {

    //  clear calibration — do this only once per calibration!
    Tet_CalibClear();

    pthread_mutex_lock(&state→mutex);
    state→status.calibstarted = 1;
    pthread_mutex_unlock(&state→mutex);
  }

  if ( state→status.calibstarted) {

    //  signal gui to draw calib point
    daimon→signal_gui(REQ_WRITE,sync,x,y);

    //  get eye tracker to start recording at this point
    Tet_CalibAddPoint(x,y,samples,gazeDataReceiver,NULL,0);

    //  block for gui to finish drawing
    if(sync) daimon→block_et();

    //  advance to next calibration point
    pthread_mutex_lock(&state→mutex);
    state→calpoint++;
    pthread_mutex_unlock(&state→mutex);

    if (  last calibration point  )  {

      //  signal gui to exit calibration
      daimon→signal_gui(REQ_QUIT,false,-1.0,-1.0);

      pthread_mutex_lock(&state→mutex);
      state→status.calibstarted = 0;
      state→status.calibrating = 0;
      state→calpoint = 0;
      pthread_mutex_unlock(&state→mutex);

      //  finished calibrating, tell the eye tracker we're done
      Tet_CalibCalculateAndSet();

      //  hint: should perform a quality check here and save calibration
      //  file for future use if calibration is good enough
    }
  }
}
```

Listing 11.1. Tobii thread: calibrate with conditional waits.

```
void main_loop(void)
{
  // clear screen
  ...

  // view transformation
  ...

  // object transformation
  ...

  if (state→status.connected) {
    if (state→status.calibrating &&
        state→status.calibstarted) {

      // block until tobii thread advances to new point
      if(daimon→block_gui(&x,&y)) {
        // draw calibration dot at (x,y) coordinates
      }

      // signal tobii thread to advance to new point
      daimon→signal_et();

    }
    else if (state→status.running) {
      // draw left gaze point (normalized coordinates scaled to window size)
      draw_point(state→xl * w, h−state→yl * h);

      // draw right gaze point
      draw_point(state→xr * w, h−state→yr * h);

      // draw average gaze point
      x = (state→xl+state→xr)/2 * w;
      y = (state→yl+state→yr)/2 * h;
      draw_point(x,h − y);

      // draw current camera eye point (left eye, right eye, average of both)
      ...
    }
  }
  glutSwapBuffers();
}
```

Listing 11.2. GUI main loop with conditional waits.

The daimon, with its C++ interface given in Listing 11.3, was modeled after Buten-hof's server, which was designed to process data requests either synchronously or asynchronously. The daimon is only used synchronously by the ET thread, hence some of the original asynchronous functionality may be missing.

```cpp
typedef enum { REQ_READ, REQ_WRITE, REQ_QUIT } operation_t;

class Request {
        public:
        Request(operation_t op, float ix, float iy) : \
                operation(op), x(ix), y(iy) { }

        class Daimon;                                            // friends

        operation_t             operation;              // function code
        float                   x, y;
};

class Daimon {
        public:
        Daimon() : \
                synchronous(false), \
                done_flag(false), \
                request_flag(false) { pthread_mutex_init(&mutex,NULL);
                                      pthread_cond_init(&request,NULL);
                                      pthread_cond_init(&done,NULL);
                                    };

        void signal_gui(operation_t operation, bool sync, float x, float y);
        bool block_gui(float *x, float *y);
        void signal_et(void);
        void block_et(void);

        private:
        deque<Request *>        queue;
        bool                    synchronous;    // true if synchronous
        bool                    done_flag;      // predicate for wait
        bool                    request_flag;   // predicate for wait
        pthread_mutex_t         mutex;
        pthread_cond_t          request;
        pthread_cond_t          done;           // wait for completion
};
```

Listing 11.3. Calibration daimon interface.

Data requests are received by the **daimon** as either read, write, or quit operations, along with calibration coordinates (x, y), embodied in object **request**. In its original server instantiation, read requests would be used to read the data that the server had finished processing. In the present case, only the write and quit operations are used by the ET thread (technically, the GUI thread reads the (x, y) coordinates from the **daimon**, but it does so implicitly only when the ET thread has written data). Requests are pushed onto a queue for retrieval by the GUI thread. The queue maintained by the **daimon** acts as a type of conduit for calibration coordinates. Using First-In-First-Out (FIFO) order ensures that coordinates are processed in the proper sequence.

Listing 11.4 lists the GUI blocking and signaling calls. Recall that the GUI thread blocks itself and is signaled by the ET thread. When blocking itself, the GUI thread waits on the conditional variable **request**. Conditional variables are usually used for communicating information about the state of shared data, e.g., signaling when a queue is no longer empty, as in the present instance. When a request is made by the ET thread, and a set of calibration coordinates is placed onto the shared queue, the GUI thread is signaled, waking it up to retrieve the coordinates and draw the calibration dot. There are two important subtleties within the **block_gui** routine:

1. The **mutex** is locked upon function entry, but is only released upon exit from the **signal_et** function. Both **block_gui** and **signal_et** calls are made by the GUI thread and should be read as one long function that is broken up by the GUI thread's act of drawing a calibration dot. The **mutex** is kept in its locked state across both functions so that the GUI thread can signal the ET thread via the **done** conditional variable.

2. The **request_flag** is the **request** conditional variable's predicate and is used in the condition variable wait loop. Waiting for a condition variable in a loop protects against multiprocessor races and spurious wakeups (see Butenhof (1997) for details). However, spurious wakeups (e.g., a missed signal) may still occur or the **request_flag** may be set inadvertently. If so, and without any further testing, the GUI thread could possibly attempt to read data from an empty queue. To prevent this, the **request_flag** is tested again as is the state of the queue prior to data retrieval.

Listing 11.5 lists the ET thread blocking and signaling calls. Recall that the ET thread blocks itself and is signaled by the GUI thread. Function **signal_et** should be considered as the tail end of **block_gui**; it is used by the GUI thread to signal (wake up) the waiting ET thread after finishing drawing of the calibration dot. Function **block_et** is what the ET thread calls to block itself. It is roughly symmetrical to **block_gui** in that a condition variable wait loop is used along with a predicate (**done_flag**) linked to the condition variable (**done**).

```cpp
void Daimon::signal_gui(operation_t operation, bool sync, float x, float y)
{
        Request      *req;

  pthread_mutex_lock(&mutex);

  synchronous = sync;

  // add new request to queue
  queue.push_back(new Request(operation,x,y));

  request_flag = true;

  // tell daimon a request is available
  pthread_cond_signal(&request);

  pthread_mutex_unlock(&mutex);
}

bool Daimon::block_gui(float *x, float *y)
{
        Request*     req=NULL;
        bool         gotdata=false;

  pthread_mutex_lock(&mutex);

  // wait for data
  while(!request_flag) pthread_cond_wait(&request, &mutex);

  // it's possible that we missed signal OR request_flag was true
  // to begin with—check queue before attempting to process the data
  if (request_flag && !queue.empty()) {
    req = queue.front(); queue.clear();                 // strip from queue
    switch(req→operation) {
      case REQ_QUIT:                                    break;
      case REQ_READ:                                    break;
      case REQ_WRITE: *x = req→x; *y = req→y; gotdata = true; break;
      default:                                          break;
    }
    if (req) delete req;                                // prevent memory leak
  }
  request_flag = false;

  return(gotdata);
}
```

Listing 11.4. Calibration daimon implementation: blocking and signaling the GUI thread.

```
void Daimon::signal_et(void)
{
  if (synchronous) {
    done_flag = true;
    pthread_cond_signal(&done);
  }
  pthread_mutex_unlock(&mutex);
}

void Daimon::block_et(void)
{
  pthread_mutex_lock(&mutex);
  while(!done_flag) pthread_cond_wait(&done,&mutex);
  done_flag = false;
  pthread_mutex_unlock(&mutex);
}
```

Listing 11.5. Calibration daimon implementation: blocking and signaling the ET thread.

11.2 Summary and Further Reading

This chapter focused on a subtle calibration problem: synchronization of drawing and tracker sampling at successive calibration coordinates in the calibration sequence. The main approach advocated here is to use concurrent process synchronization, namely conditional waits. The use of standardized POSIX threads is suggested, with Butenhof (1997) acting as a suitable text for learning various intricacies of thread programming. Other concurrent programming texts may also be suitable.

Of course, it must again be stated that the simpler form of concurrency via bitfields or flag variables is probably sufficient to achieve good calibration, particularly if proper verbal instructions are given to users.

Beyond the synchronization issue, most of the points of calibration given in Chapter 8 still apply to the table-mounted system. Specifically, it is still important to choose good calibration coordinates and to check calibration drift and error.

Calibration coordinates should be selected to cover the extents of the viewing area. The use of five calibration points at the corners and center is a good strategy, although tracker manufacturers are beginning to suggest the use of as few as two points at the corners. As technology and computer vision algorithms improve, calibration may become obsolete altogether and make way for autocalibrating eye trackers. However, this advancement is still a few years away. For the time being, if accuracy is important to the given eye tracking application, drift and error should still be checked (and reported in eventual results reports or papers).

Calibration error checking can be performed via a strategy similar to the one given in Section 8.2.1. This consists of measuring the average distance (error) between the known calibration point coordinate and the average distance of the multiple gaze samples obtained by the eye tracker during calibration sampling. Modern eye trackers, such as the Tobii, facilitate this during calibration by providing validity data obtained during calibration. For example, Tobii sends the **STet_CalibAnalyzeData** data structure to the eye tracking application that contains information suitable for calibration error checking. This can be used to measure the error at specific calibration coordinates or provide visual feedback display regarding the "goodness of fit" of the calibration. These indicators can then be used to either recalibrate specific coordinates or to quantify the calibration.

12

Eye Movement Analysis

The goal of eye movement measurement and analysis is to gain insight into the viewer's attentive behavior. As can be seen in Figure 8.5 at the end of Chapter 8, raw eye movement data, or perhaps data processed to a certain extent such as Gaze Intersection Point (GIP) data in virtual reality, may appear to be informative, however, without further analysis, raw data are for the most part meaningless. Although intuitively (and from the knowledge of the task), it is possible to guess where the subject happened to be paying attention in the environment (over the internal calibration points, as she or he was instructed), it is not possible to make any further quantitative inferences about the eye movement data without further analysis. A method is needed to identify fixations; those eye movements that best indicate the locations of the viewer's (overt) visual attention.

As suggested in Chapter 4, eye movement signals can be approximated by linear filters. Fixations and pursuits can be modeled by a relatively simple neuronal feedback system. In the case of fixations, the neuronal control system is responsible for minimizing fixation error. For pursuit movements, the error is similarly measured as distance off the target, but in this case the target is nonstationary. Fixations and pursuits may be detected by a simple linear model based on linear summation.

The linear approach to eye movement modeling is an operational simplification of the underlying nonlinear natural processes (Carpenter, 1977). The linear model assumes that position and velocity are processed by the same neuronal mechanism. The visual system processes these quantities in different ways. The position of a target is signaled by the activation of specific retinal receptors. The velocity of the target, on the other hand, is registered by the firing rate (amplitude) of the firing receptors. Furthermore, nonlinearities are expected in most types of eye movements. Accelerational and decelerational considerations alone suggest the inadequacy of the linear assumption. Nevertheless, from a signal processing standpoint, linear filter analysis is sufficient for the localization of distinct features in eye movement signals. Although this approach is a poor estimate of the underlying system, it nonetheless establishes a useful approximation of the signal in the sense of pattern recognition.

The goal of eye movement signal analysis is to characterize the signal in terms of salient eye movements, i.e., saccades and fixations (and possibly smooth pursuits). Typically, the analysis task is to locate regions where the signal average changes abruptly indicating the end of a fixation and the onset of a saccade and then again assumes a stationary characteristic indicating the beginning of a new fixation. A hypothetical plot of an eye movement time course is shown in Figure 12.1. The graph shows the sought points in the signal where a saccade begins and ends. Essentially, saccades can be thought of as signal edges in time.

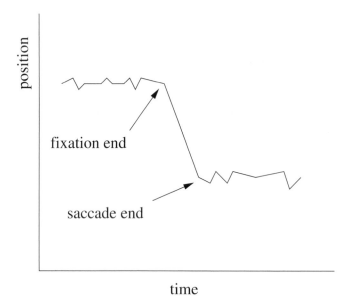

Fig. 12.1. Hypothetical eye movement signal.

Two main automatic types of approaches have been used to analyze eye movements: one based on summation (averaging), the other on differentiation.[1] In the first, the temporal signal is averaged over time. If little or no variance is found, the signal is deemed a candidate for fixation. Furthermore, the signal is classified as a fixation, provided the duration of the stationary signal exceeds a predetermined threshold. This is the "dwell-time" method of fixation determination. In the second, assuming the eye movement signal is recorded at a uniform sampling rate, successive samples

---

[1] A third type of analysis, requiring manual intervention, relies on slowly displaying the time course of the signal (the scanpath), either in 1D or 2D, one sample at a time, and judging which sample points lie outside the mean. This "direct inspection" method is rather tedious, but surprisingly effective.

are subtracted to estimate eye movement velocity. The latter type of analysis is gaining favor, and appears more suitable for real-time detection of saccades. Fixations are either implicitly detected as the portion of the signal between saccades, or the portion of the signal where the velocity falls below a threshold.

Thresholds for both summation and differentiation methods are typically obtained from empirical measurements. The seminal work of Yarbus (1967) is often still referenced as the source of these measurements.

12.1 Signal Denoising

Before (or during) signal analysis, care must be taken to eliminate excessive noise in the eye movement signal. Inevitably, noise will be registered due to the inherent instability of the eye, and worse, due to blinks. The latter, considered to be a rather significant nuisance, generates a strong signal perturbation, which (luckily) may often be eliminated, depending on the performance characteristics of the available eye movement recoding device. It is often the case that either the device itself has capabilities for filtering out blinks, or that it simply returns a value of $(0,0)$ when the eye tracker "loses sight" of the salient features needed to record eye movements.

In practice, eye movement data falling outside a given rectangular range can be considered noise and eliminated. Using a rectangular region to denoise the (2D) signal also addresses another current limitation of eye tracking devices: their accuracy typically degrades in extreme peripheral regions. For this reason (as well as elimination of blinks), it may be sensible to simply ignore eye movement data falling outside the "effective operating range" of the device. This range will often be specified by the vendor in terms of visual angle. An example of signal denoising is shown in Figure 12.2, where an interior rectangular region 10 pixels within the image borders defines the operating range. Samples falling outside this constrained interior image boundary were removed from the record (the original images measured 600×450 pixels).

12.2 Dwell-Time Fixation Detection

The dwell-time fixation detection algorithm depends on two characterization criteria:

1. Identification of a stationary signal (the fixation)
2. Size of time window specifying an acceptable range (and hence temporal threshold) for fixation duration

An example of such an automatic saccade/fixation classification algorithm, suggested by Anliker (1976), determines whether M of N points lie within a certain distance D of the mean (μ) of the signal. This strategy is illustrated in Figure 12.3a where two N-sized windows are shown. In the second segment (positioned over a

(a) Raw eye movement data.

(b) Eye movement data after eliminating samples falling outside an interior region ten pixels inside the image borders.

Fig. 12.2. Eye movement signal denoising. Courtesy of Wesley Hix, Becky Morley, and Jeffrey Valdez. Reproduced with permission, Clemson University.

hypothetical saccade), the variance of the signal would exceed the threshold D indicating a rapid positional change, i.e., a saccade. The values of M, N, and D are determined empirically. Note that the value of N defines an a priori sliding window of sample times where the means and variances are computed. Anliker denotes this algorithm as the *position-variance method* because it is based on the fact that a fixation is characterized by relative immobility (low position variance) whereas a saccade is distinguished by rapid change of position (high position variance).

12.3 Velocity-Based Saccade Detection

An alternative to the position-variance method is the *velocity detection method* (Anliker, 1976). In this velocity-based approach, the velocity of the signal is calculated within a sample window and compared to a velocity threshold. If the sampled velocity is smaller than the given threshold, then the sample window is deemed to belong to a fixation signal, otherwise it is a saccade. The velocity threshold is specified empirically. Figure 12.3b shows the hypothetical eye movement time course with the sample window centered over the saccade with its velocity profile above.

Noting Yarbus' observation that saccadic velocity is nearly symmetrical (resembling a bell curve), a velocity-based prediction scheme can be implemented to approximate the arrival time and location of the next fixation. The next fixation location can be approximated as soon as the peak velocity is detected. Measuring elapsed time and distance traveled, and taking into account the direction of the saccade, the prediction scheme essentially mirrors the left half of the velocity profile (up to its peak) to calculate the saccade's end point.

The position-variance and velocity-based algorithms give similar results, and both methods can be combined to bolster the analysis by checking for agreement. The velocity-based algorithm offers a slight advantage in that often short-term differential filters can be used to detect saccade onset, decreasing the effective sample window size. This speeds up calculation and is therefore more suitable for real-time applications. The image in Figure 12.2b was processed by examining the velocity,

$$v = \frac{\sqrt{(x_{t+1} - x_t)^2 + (y_{t+1} - y_t)^2}}{dt},$$

between successive sample pairs and normalizing against the maximum velocity found in the entire (short) time sequence. The normalized value, subtracted from unity, was then used to shade the sample points in the diagram. Slow moving points, or points approaching the speed of fixations are shown brighter than their faster counterparts.

Tole and Young (1981) suggest the use of short FIR (Finite Impulse Response) filters for saccade and fixation detection matching idealized saccade signal profiles. An idealized (discrete) saccade time course is shown in Figure 12.4a. A sketch of

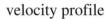

(a) Position-variance method.

velocity profile

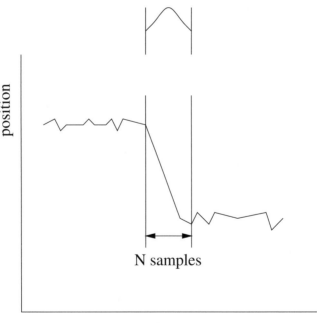

(b) Velocity-detection method.

Fig. 12.3. Saccade/fixation detection.

the algorithm proposed by Tole and Young is presented in Figure 12.4b, with corresponding FIR filters given in Figure 12.5. The algorithm relies on four conditions to detect a saccade:

$$|I_1| > A \qquad (12.1)$$
$$|I_2| > B \qquad (12.2)$$
$$Sgn(I_2) \neq Sgn(I_1) \qquad (12.3)$$
$$T_{min} < I_2 - I_1 < T_{max}. \qquad (12.4)$$

If the measured acceleration exceeds a threshold (condition (12.1)), the acceleration buffer is scanned forward to check for a second peak with opposite sign to that of I_1 (condition (12.3)) and greater in magnitude than threshold B (condition (12.2)). Amplitude thresholds are derived from theoretical values, e.g., corresponding to expected peak saccade velocities of 600°/s (visual angle). Condition (12.4) stipulates minimum and maximum durations for the forward search, also based on theoretical limits, e.g., saccade durations in the range 120 ms–300 ms. If all conditions are met, a saccade is detected.

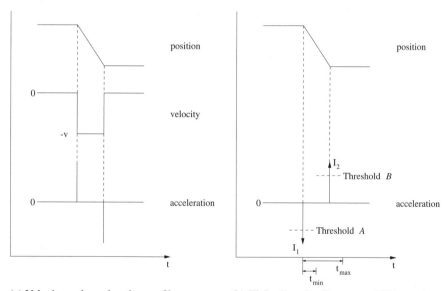

(a) Velocity and acceleration profiles.

(b) Finite Impulse Response (FIR) acceleration filter algorithm.

Fig. 12.4. Idealized saccade detection.

Enhancements to the above algorithm, offered by Tole and Young, include adaptive threshold estimation, adjusted to compensate for recently observed signal noise. This is achieved by reformulating the thresholds A and B as functions of an RMS estimate

-1	-1	0	1	2	1	0	-1	-1

acceleration

-3	-2	-1	0	1	2	3

velocity

Fig. 12.5. Finite Impulse Response (FIR) filters for saccade detection.

of noise in the signal:

$$\text{Threshold } A = 4000 \text{ deg/sec}^2 + \text{Accel/RMS} + \text{Accel/DC noise}$$
$$\text{Threshold } B = 4000 \text{ deg/sec}^2 + \text{Accel/RMS}.$$

The term

$$\text{Accel/DC noise} = \frac{2 \,|\, \Delta p \,|}{\Delta t^2}$$

estimates noise in acceleration, where Δt is the sampling interval and Δp is the peak-to-peak estimate of noise in position. This term can be estimated by measuring the peak-to-peak amplitude of fluctuations on the input signal when the average eye velocity over the last 2 seconds is less than 4 deg/sec. This estimate can be calculated once, for example, during calibration when the signal-to-noise ratio could be expected to remain constant over a test session, e.g., when the subject is known to be fixating a test target. This term may also be updated dynamically whenever velocity is low. The remaining estimate of noise in the output filter,

$$\text{Accel/RMS} = \sqrt{\frac{1}{T} \sum_{i=1}^{T} (\text{Accel}^2(i) - \overline{\text{Accel}}^{\,2})},$$

assumes mean acceleration is zero for time windows greater than 4 seconds; i.e., $\overline{\text{Accel}} \to 0$ as $T > 4$ s. The noise estimation term can be further simplified to avoid the square root,

$$\text{Accel/RMS} \leq \frac{1}{T} \sum_{i=0}^{T} |\,\text{Accel}(i)\,|, \quad T > 4 \text{ s},$$

because $\sqrt{\sum \text{Accel}^2} \leq \sum |\,\text{Accel}\,|$. This adaptive saccade detection method is reported to respond well to a temporarily induced increase in noise level, e.g., when the subject is asked to grit his or her teeth.

12.4 Eye Movement Analysis in Three Dimensions

Analysis of eye movements in 3D, as recorded in virtual reality, for example, follows the above algorithms, with slight modifications. In general, there are two important considerations: elimination of noise and identification of fixations. Noise may be present in the signal due to eye blinks, or other causes for the eye tracker's loss of proper imaging of the eye. These types of gross noise artifacts can generally be eliminated by knowing the device characteristics, for example, the eye tracker outputs $(0,0)$ for eye blinks or other temporary missed readings. Currently there are two general methods for identification of fixations: the position-variance strategy or the velocity-detection method. A third alternative may be a hybrid approach that compares the result of both algorithms for agreement.

The traditional two-dimensional eye movement analysis approach starts by measuring the visual angle of the object under inspection between a pair (or more) of raw eye movement data points in the time series (i.e., the POR data denoted (x_i, y_i)). Given the distance between successive POR data points, $r = \| (x_i, y_i), (x_j, y_j) \|$, the visual angle θ is calculated by the equation: $\theta = 2\tan^{-1}(r/2D)$, where D is the (perpendicular) distance from the eyes to the viewing plane. Note that r and D, expressed in like units (e.g., pixels or inches), are dependent on the resolution of the screen on which the POR data was recorded. A conversion factor is usually required to convert one measure to the other (e.g., screen resolution in dots per inch (dpi) converting D to pixels). The visual angle θ and the difference in timestamps Δt between the POR data points allows velocity-based analysis, because $\theta/\Delta t$ gives eye movement velocity in degrees visual angle per second.

Note that the arctangent approach assumes that D is measured along the line of sight, which is assumed to be perpendicular to the viewing plane. Traditional 2D eye movement analysis methods can therefore be applied directly to raw POR data in the eye tracker reference frame. As a result, identified fixations could then be mapped to world coordinates to locate fixated ROIs within the virtual environment. A different approach may be followed by mapping raw POR data to world coordinates first, followed by eye movement analysis in three-space. The main difference of this approach is that calculated gaze points in three-space provide a composite three-dimensional representation of both the user's left and right eye movements. Applying the traditional 2D approach prior to mapping to (virtual) world coordinates suggests a componentwise analysis of left and right eye movements (in the eye tracker's reference frame) possibly ignoring depth. In three dimensions, depth information is implicitly taken into account prior to analysis. However, the assumption of a perpendicular visual target plane does not hold because the head is free to translate and rotate within six degrees of freedom.

Operating directly on Gaze Intersection Point (GIP) data in (virtual) world coordinates (see Chapter 7), a fixation detection algorithm based on estimation of velocity proceeds as follows. Given raw gaze intersection points in three dimensions, the velocity-based thresholding calculation is in principle identical to the traditional 2D approach, with the following important distinctions.

1. The head position **h** must be recorded to facilitate the calculation of the visual angle.
2. Given two successive GIP data points in three-space $\mathbf{p}_i = (x_i, y_i, z_i)$ and $\mathbf{p}_{i+1} = (x_{i+1}, y_{i+1}, z_{i+1})$, and the head position at each instance $\mathbf{h}_i$ and $\mathbf{h}_{i+1}$, the estimate of instantaneous visual angle at each sample position θ_i is calculated from the dot product of the two gaze vectors defined by the difference of the gaze intersection points and averaged head position:

$$\theta_i = \cos^{-1} \frac{\mathbf{v}_i \cdot \mathbf{v}_{i+1}}{\| \mathbf{v}_i \| \| \mathbf{v}_{i+1} \|}, \ i \in [0, n), \tag{12.5}$$

where n is the sample size and $\mathbf{v}_i = \mathbf{p}_i - \overline{\mathbf{h}}$ and $\overline{\mathbf{h}}$ is the averaged head position over the sample time period. Head position is averaged because the eyes can accelerate to reach a target fixation point much more quickly than the head (Watson et al., 1997).

With visual angle θ_i and timestamp difference between $\mathbf{p}_i$ and $\mathbf{p}_{i+1}$, the same velocity-based thresholding is used as in the traditional 2D case. No conversion between screen resolution and distance to target is necessary because all calculations are performed in world coordinates.

The algorithm generalizes to the use of wider filters (by changing the subscript $i + 1$ to $i + k$ for $k > 1$) for improved smoothing. Using Equation (12.5) to calculate θ_i, only two successive data points are used to calculate eye movement velocity. This is analogous to the calculation of velocity using a two-tap FIR filter with coefficients $\{1, 1\}$.

To address excessive noise in the eye movement signal the two-tap FIR filter can be replaced by a five-tap FIR filter, as shown in Figure 12.7a. Due to its longer sampling window, the 5-tap filter is more effective at signal smoothing (antialiasing). Following Tole and Young's (1981) work, an acceleration filter may also be used on 3D GIP data, with slight modification. The acceleration filter is shown in Figure 12.7(b), and is convolved with eye movement velocity data as obtained via either the two-tap or five-tap velocity filter. The filter responses resemble the real velocity and acceleration curves for a saccade characterized in Figure 12.6.

The 3D fixation eye movement analysis algorithm calculates the velocity and acceleration at each instantaneous estimate of visual angle θ_i. Note that θ_i is effectively a measure of instantaneous eye movement magnitude (i.e., amplitude), and therefore implicitly represents eye movement velocity. That is, the signal resembles the positively oriented velocity peaks shown in Figure 12.6b. Withholding division by the time difference between successive samples (Δt) facilitates the measurement of velocity with arbitrarily long filters.

Velocity is obtained via convolution with pattern-matching FIR filters of variable length. When convolved, these filters respond to sampled data with profiles matching

(a) Position.

(b) Velocity.

(c) Acceleration.

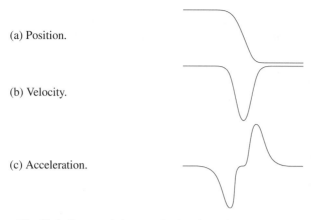

Fig. 12.6. Characteristic saccade signal and filter responses.

those of the filter. These filters, denoted $\mathbf{h}_k$, are essentially unnormalized low-pass filters which tend to smooth and amplify the underlying signal. Division by the duration of the sampling window yields velocity; i.e.,

$$\dot{\theta}_i = \frac{1}{\Delta t} \sum_{j=0}^{k} \theta_{i+j} \mathbf{h}_j, \ i \in [0, n-k),$$

expressed in deg/s, where k is the filter length, $\Delta t = k - i$.

Acceleration is obtained via a subsequent convolution of velocity $\dot{\theta}_i$, with the acceleration filter $\mathbf{g}_j$ shown in Figure 12.7b. That is,

$$\ddot{\theta}_i = \frac{1}{\Delta t} \sum_{j=0}^{k} \dot{\theta}_{i+j} \mathbf{g}_j, \ i \in [0, n-k),$$

where k is the filter length, $\Delta t = k - i$. The acceleration filter is essentially an unnormalized high-pass differential filter. The resulting value $\ddot{\theta}_i$ expressed in deg/s^2, is checked against threshold A. If the absolute value of $\ddot{\theta}_i$ is greater than A, then the corresponding gaze intersection point $\mathbf{p}_i$ is treated as the beginning of a saccade. Scanning ahead in the convolved acceleration data, each subsequent point is tested in a similar fashion against threshold B to detect the end of the saccade. Two additional conditions are evaluated to locate a saccade, as given by Tole and Young. The four conditions are listed and illustrated in Figure 12.8.

Note that these velocity and acceleration filters differ from those used by Tole and Young. This is because Tole and Young applied their filters (the reverse of these, essentially) to the positional eye movement signal ($\mathbf{p}$), whereas the filters given here are applied to the signal amplitude (θ). Pseudocode of the technique is presented in Listing 12.1.

```
// Input: p(n), gaze intersection points, h(k), g(k), velocity and accel. filters, resp.
// Output: classification of each pᵢ as fixation or saccade

// convert GIP data to angular GIP data (instantaneous visual angle)
for(i = 0 to n − 1)
   θᵢ = cos⁻¹(vᵢ · vᵢ₊₁ / ‖ vᵢ ‖‖ vᵢ₊₁ ‖);

// init accumulation arrays (convolution results)
for(i = 0 to n − k − 1)
   θ̇ᵢ = θ̈ᵢ = 0;

// convolve (vel)
for(i = 0 to n − k − 1)
   for(j = 0 to k)
      θ̇ᵢ = θ̇ᵢ + θᵢ₊ⱼhⱼ;

// convolve (accel)
for(i = 0 to n − k − 1)
   for(j = 0 to k)
      θ̈ᵢ = θ̈ᵢ + θ̇ᵢ₊ⱼgⱼ;

// find all saccadic intervals
for(i = 0 to n − k − 1) {
   if (| θ̈ᵢ | ≥ A) {
      // (condition 1)
      for(j = i+Tₘᵢₙ to j < (n − k) − i && (j − i) ≤ Tₘₐₓ) {
         // (condition 4 implicit in loop)
         if (| θ̈ᵢ₊ⱼ | ≥ B && sgn(θ̈ᵢ₊ⱼ) ≠ sgn(θ̈ᵢ)) {
            // (conditions 2 & 3)
            for(l = i to l < j)
               pₗ = saccade
         } else {
            pᵢ = fixation
         }
      }
   }
}
```

Listing 12.1. Acceleration-based saccade detection.

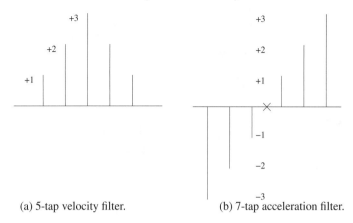

(a) 5-tap velocity filter.   (b) 7-tap acceleration filter.

Fig. 12.7. FIR filters.

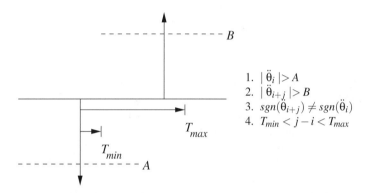

1. $|\ddot{\theta}_i| > A$
2. $|\ddot{\theta}_{i+j}| > B$
3. $sgn(\ddot{\theta}_{i+j}) \neq sgn(\ddot{\theta}_i)$
4. $T_{min} < j - i < T_{max}$

Fig. 12.8. Acceleration thresholding.

12.4.1 Parameter Estimation

Thresholds are needed for saccade velocity, acceleration, and duration, because the fixation detection algorithm relies on the detection of saccades. Although algorithm parameters may eventually be determined empirically, algorithm fine tuning is guided by a review of the literature, briefly summarized here for context.

The duration of saccades is related in a nonlinear manner to their amplitude over a thousandfold range ($3'$–$50°$) (Bahill et al., 1975). Saccades of less than 15 or 20 degrees in magnitude are physiologically the most important because most naturally occurring saccades fall in this region. When looking at pictures, normal scanpaths are characterized by a number of saccades similar in amplitude to those exhibited during reading. The saccade "main sequence" describes the relationships between saccade duration, peak velocity, and magnitude (amplitude). Because saccades are generally stereotyped, the relationship between saccade amplitude and duration can be modeled by the linear equation $\Delta t = 2.2\theta + 21$ (Knox, 2001). Peak velocity reaches a soft saturation limit up to about 15 or 20 degrees, but can range up to about 50°, reaching velocity saturation at about 1000 deg/s (Clark & Stark, 1975). In practice, the main sequence relationship between amplitude and velocity can be modeled by the asymptotic equation $\dot{\theta} = \lambda(1 - e^{-\theta/15})$, with velocity upper limit (asymptote λ) set to 750 deg/s (Hain, 1999).

According to accepted saccade amplitude estimates, measured instantaneous eye movement amplitude (θ) is expected to range up to about 20°. Example data captured in VR ranges up to 136°, with mean 1.5° (median 0.27°) and 9.7° s.d., which appears to be within normal limits, except for a few outliers (possibly due to head motion; see Figure 12.9a).

For saccade detection via velocity filtering, a threshold of 130 deg/s may be chosen for both two-tap and five-tap filters. Using the asymptotic model of the main sequence relationship between saccade amplitude and velocity (limited by 750 deg/s), this threshold should effectively detect saccades of amplitude roughly greater than 3°. Observed velocity averages are reported in Table 12.1 (see also Figures 12.9(b) and 12.9(c)).

Saccade detection via acceleration filtering requires setting a larger number of parameters. Values of 10 ms and 300 ms for T_{min} and T_{max}, respectively, appear to cover a fairly wide range of saccade acceleration impulse pairs. The choice of the remaining threshold for saccade acceleration is made difficult because no applicable models of saccadic acceleration (e.g., a main sequence) could readily be found. In fact, unlike commonly listed limits of amplitude, duration, and velocity, there seems to be some disagreement regarding upper limits of acceleration. Peak acceleration has been reported to average at about 30,000 deg/s² in saccades of 10° with a saturation limit of 35,000 deg/s² for $\theta < 15°$, whereas other findings are given of 20° saccades with average peak acceleration of 26,000 deg/s² (Becker, 1989). Observed acceleration averages are reported in Tables 12.1 and 12.2 (see also Figure 12.9d). Following Tole and Young's acceleration filtering algorithm (incidentally, these authors report acceleration limits approaching 80,000 deg/s²), the authors' recommended thresholds for saccade acceleration are suitable initial estimates (see below).

In Tole and Young's paper the authors point out variable noise characteristics dependent on the subject's actions (e.g., different noise profile while gritting teeth). To adapt to such signal changes the authors recommend an adaptive thresholding technique that dynamically adjusts the threshold, based on the current estimate of noise level. Indeed, a very large peak-to-peak acceleration signal variance may be observed (see Figure 12.9d). As the authors suggest, an adaptive thresholding technique may aid in automatically setting acceleration thresholds A and B:

$$A = B = 1000 + \sqrt{\frac{1}{k}\sum_{i=0}^{k}(\ddot{\theta}_{i+k})^2} \text{ deg/s}^2,$$

(a) Eye movement amplitude θ (deg. visual angle)

(b) Velocity $\dot{\theta}$ (deg/s) via 2-tap filter; saccade indicator

(c) Velocity $\dot{\theta}$ (deg/s) via 5-tap filter; saccade indicator

(d) Accel. $\ddot{\theta}$ (deg/s^2); threshold; saccade indicator

Fig. 12.9. Eye movement signal and filter responses.

Table 12.1. Velocity algorithm comparisons.

	2-Tap	5-Tap
Fixation groups	30	21
Mean fixation duration (ms)	1079	1450
Time spent in fixations	73%	69%
Min θ (deg/s)	0	.917
Max θ (deg/s)	12,385	5,592
Avg. θ (deg/s)	106	106
S.d. θ (deg/s)	635	451

Table 12.2. Acceleration algorithm comparisons.

	2-Tap		5-Tap	
Thresholding	Adaptive	Constant	Adaptive	Constant
Fixation groups	20	17	17	14
Mean fixation duration (ms)	1633	1583	1937	1983
Time spent in fixations	74%	61%	74%	63%
Min $\ddot{\theta}$ (deg/s^2)		-257,653		-182,037
Max $\ddot{\theta}$ (deg/s^2)		248,265		167,144
Avg. $\ddot{\theta}$ (deg/s^2)		4,453		3,966
S.d. $\ddot{\theta}$ (deg/s^2)		22,475		17,470

where k is the number of samples in time T proportional to the length of the acceleration filter; that is,

$$T = \frac{\text{filter length}}{\text{sampling rate}} = \frac{9}{30\ \text{Hz}} = 300\ \text{ms}.$$

This is a slightly different implementation of adaptive thresholding than Tole and Young's. The threshold value is slightly lower and its adaptive adjustment relies on explicit calculation of the acceleration Root Mean Squared (RMS). Also, the sampling window for this purpose is much shorter than the authors' recommended window of $T > 4$ s. Finally, the adaptive technique given above employs a "lookahead" scan of the acceleration data, suitable for off-line analysis. Changing the $i + k$ subscript to $i - k$ provides a "lookbehind" scan that can be employed in real-time systems.

12.4.2 Fixation Grouping

The above algorithm classifies each GIP as either part of a fixation or saccade (see the saccade indicator plots in Figure 12.9). Once each GIP has been classified, each string of consecutive fixation GIPs is condensed to a single fixation point by finding the centroid of the group. However, due to the nature of the new algorithm, we observed that at times isolated noisy GIPs were also included in fixation groups. To prevent the inclusion of such outlying points a simple check can be implemented to verify that each fixation group's duration is greater than or equal to the minimum theoretical fixation duration (i.e., 150 ms; Irwin (1992)). Augmenting the acceleration-based saccade detection described above, this check can be considered as a position-variance component (emphasizing temporal coherence over spatial distribution) of a hybridized spatiotemporal approach to eye movement signal analysis.

12.4.3 Eye Movement Data Mirroring

Although the 3D eye movement analysis algorithm is mathematically robust at handling signal noise, the system is still susceptible to noise generated by the eye tracker. In particular, eye tracking equipment may randomly drop POR data. In some cases (e.g., during a blink), null POR values may be recorded for both left and right eyes. However, in some instances, only one eye's POR may be null while the other is not. This may occur due to calibration errors. To address this problem a heuristic technique may be used to mirror the nonnull POR eye movement data (Duchowski et al., 2002). The table in Figure 12.10 shows an example of this technique. The left eye POR at time $t + 1$ is recorded as an invalid null point. To estimate a nonnull left eye coordinate at $t + 1$, the difference between successive right eye POR values is calculated and used to update the left eye POR values at $t + 1$, as shown in the equation in Figure 12.10, giving $(x_{l_{t+1}}, y_{l_{t+1}}) = (-0.5 + dx, 0 + dy) = (-0.4, 0)$. Note that this solution assumes static vergence eye movements. It is assumed that the eyes remain at a fixed interocular distance during movement. That is, this heuristic strategy will clearly not account for vergence eye movements occurring within the short corrective time period.

Time	Left Eye	Right Eye
t	$(-0.5, 0)$	$(0.3, 0)$
$t + 1$	$(0, 0)$	$(0.4, 0)$

$$dx = x_{r_{t+1}} - x_{r_t} = 0.4 - 0.3 = 0.1$$
$$dy = y_{r_{t+1}} - y_{r_t} = 0.0 - 0.0 = 0.0$$

Fig. 12.10. Heuristic mirroring example and calculation.

12.5 Summary and Further Reading

Eye movement analysis for 2D applications is fairly straightforward. The 3D eye movement analysis technique is resolution-independent and is carried out in three-space, which is particularly suitable for VR applications.

There are various collections of technical papers on eye movements, usually assembled from proceedings of focused symposia or conferences. A series of such books was produced by John Senders et al. in the 1970s and 1980s (see, e.g., Monty and Senders (1976) and Fisher et al. (1981)). Good papers surveying eye movement analysis protocols have also recently appeared. Salvucci and Goldberg (2000) provide a good survey of current techniques (see also Salvucci and Anderson (2001)).

Part III

Eye Tracking Methodology

13

Experimental Design

This chapter deals with a general review of experimental design with emphasis on lab-based experiments. Generally speaking, this excludes nonexperimental designs (e.g., observational studies) and fieldwork. The point of this chapter is to review experimental procedures and methods, providing a framework, or context if you will, for eye tracking experiments. The content of this chapter reviews basic experimental designs, and does not differ greatly from texts used in introductory experimental psychology classes. For example, see Coolican (1996) for such a text. From this set of designs, this chapter identifies factorial designs as particularly popular in eye tracking research.

13.1 Formulating a Hypothesis

When designing an experiment, one of the first considerations should be the formulation of the research question. Formulating this question properly should lead the researcher toward a good design. For example, starting out with a statement such as, "I wonder what would happen if...," is what could be considered a naïve approach because it is not necessarily based on any assumptions or theories and does not identify any particular direction for testing. On the other hand, stating, "I bet this result would happen if...," already suggests an underlying assumption as well as potential candidate measures, e.g., some quantity that can be measured during experimental outcomes. The point is that a hypothesis is required when designing a formal experiment. Given a hypothesis, the experiment almost "designs itself" because it is then mainly concerned with accepting or rejecting the preliminary hypothesis, if it is stated with sufficient precision.

More formally, an experimental design is often drawn from the formulation of a *null hypothesis* (H_0), i.e., a statement predicting no difference in measured results collected between two (or more) sets of data obtained under different conditions. Hence, no effect is expected. The point of the experiment then is to reject the null hypothesis, showing that results are highly unlikely if the null hypothesis is true, thereby providing support for the alternative hypothesis. A classic example that is familiar to most people is that of a new drug being tested. The null hypothesis states that the drug has no effect, or more specifically, its effect is no different from a placebo (a sugar pill that is known not to have any effect). Establishing the hypothesis immediately suggests a logical course of action: how to administer the drug, and what to measure. The drug, or treatment, could be administered to one group of participants, with another group of participants receiving the placebo. Measurements can then be compared between the two groups. Using statistical analysis, if the measurements are no different, then the null hypothesis is accepted (indicating the new drug's inefficacy), otherwise, the null hypothesis is rejected. The latter case lends support to the statement that the drug has an effect. Note, however, that this support does not constitute absolute proof. The experiment, rooted in the conventional scientific method, provides scientific evidence for the drug's effect, but not proof of its effectiveness. This perhaps subtle distinction between scientific evidence and proof is too often ignored by students and overzealous marketing agencies.

Eye tracking studies generally do not involve pills or other digestibles. Instead, the "treatment" is often some differing form of interactive display, e.g., the computer's Graphical User Interface, or GUI (pronounced "gooey"), or varying forms of visual stimulus, e.g., two different images. In most cases, study participants are often given fairly specific tasks such as execution of some function, e.g., open a Web browser or find a specific GUI icon. Measurements then include reaction times (how quickly participants perform the action, on average), error rates (how many mistakes occurred, on average), and, of course, measures related to participants' eye movements. The latter usually include fixations, fixation durations, etc. How the different conditions are manipulated within the experiment is governed by the study's experimental design.

More formally, the treatment being manipulated or changed in value is referred to as the Independent Variable, or IV. All other variables are held constant (or attempted to be held constant; variables outside the experiment's control affecting the measured outcome may confound the outcome and are known as confounding variables). Whatever is being measured (e.g., reaction time) is usually whatever is expected to be affected by the IV, and is known as the Dependent Variable, or DV. That is, the DV depends on the manipulation of the IV.

The remainder of this chapter reviews different types of experimental inquires that can be made to test a given hypothesis. Given the choice of one design over another, the chapter then provides a review of basic statistical tools that can be used to measure differences and hence the effect of the conditions under examination.

13.2 Forms of Inquiry

Coolican (1996) defines *investigation* as a general term for any study that seeks information (usually to test a hypothesis). An *experiment* is a particular form of study where, in general, all possible causes of variation in the effect being measured are eliminated except the one influence under investigation. The general rule of thumb is to vary one thing while keeping everything else constant. Ensuring that all other conditions are equal except the main effect suggests gaining control of the experiment. This is the key concern of experimental designs: how to ensure that only one condition is varied and all else is held constant. This may sound simple but it is not. Even in fairly highly controlled settings such as laboratories, there are still many factors that may influence the outcome of experiments. Simple and mundane considerations such as whether study participants performed their given tasks before or after lunch may matter. The degree to which conditions are controllable will determine the type of experiment (or nonexperiment) being conducted.

There are a few different dimensions that specify different forms of experimental designs, including:

- Experiments versus observational studies
- Laboratory versus field research
- Idiographic versus nomothetic research
- Sample population versus single-case experiment versus the case study
- Within-subject (repeated measures) versus between-subjects designs

13.2.1 Experiments Versus Observational Studies

The distinction between experiments and nonexperimental observational studies revolves about the manipulation of an independent variable. Observational studies are generally made by observation without manipulation of an IV (e.g., consider gender as an IV; it cannot be manipulated). Being able to manipulate an IV is generally a prerequisite for the design of an experiment. Furthermore, in the interest of replicability, experiments often follow a standardized procedure. Variables, independent and dependent, need to be strictly defined, procedures undertaken during experimental trials need to be detailed, and results from analysis must be effectively reported. Most research papers follow a fairly similar format, partially so that other researchers can reproduce their experiments and (it is hoped) replicate their results. This format often includes:

1. Hypothesis: the null or alternative hypothesis, with theoretical justification for any given assumptions.
2. Design: which experimental design is ultimately chosen, is it a nonexperimental observational study, or if an experiment, what are the IVs and DVs, and how are participants grouped, if at all (e.g., within-subjects or between-subjects; see below).

3. Participants: the number of participants in the study, with demographic data such as age ranges and gender distribution (all reported anonymously).

4. Apparatus: the devices used; in eye tracking studies, one generally reports the operating characteristics of the eye tracker including its underlying mechanism (e.g., video-based, combined pupil–corneal reflection), accuracy (e.g., $0.5°$), sampling rate (e.g., 50 Hz), operating range (e.g., 50 cm), and whether any other auxiliary devices such as chin rests are needed.

5. Procedures: essentially what is told to participants prior to and following their experimental trials; is there any training or instructions (usually read from a script), what type of calibration is used, etc.

6. Tasks: what do the participants actually do? Task definition is particularly important, more so for eye tracking studies because eye movements are known to be task-dependent (gaze is simultaneously bottom-up, stimulus-driven as well as top-down, goal-oriented).

13.2.2 Laboratory Versus Field Research

Conducting an experiment in the laboratory can often allow greater control over experimental conditions than what can normally be achieved in the field. Control is probably the chief reason for holding experiments in the laboratory. Indeed, for various computer-related experiments, such as usability testing, numerous usability labs have appeared with specialized recording "studios" equipped with one-way mirrors, video cameras, and eye trackers. Detractors of lab experiments question the generalizability of results to less artificial settings of one's office, home, etc. In a nutshell, laboratory experiments suffer from a reduction of ecological validity but, through increased control, gain internal validity.

For eye tracking research, equipment often dictates pragmatic constraints such as whether the experiment needs to remain in the lab or whether the eye tracker can be used out "in the field". With increasingly smaller and more portable equipment, eye tracking experiments need not be confined to the lab. For example, table-mounted eye tracking equipment can be fairly easily transported and with a laptop experiments can be conducted "on-site". Head-mounted gear is also becoming increasingly less cumbersome and more affordable (Li et al., 2006) and hence can be used for various experiments performed outside the lab.

13.2.3 Idiographic Versus Nomothetic Research

This distinction pertains to the study of an individual (idiographic) versus the study of larger populations. Generally speaking, beyond clinical evaluations of individuals, or evaluation of custom-built solutions, eye tracking studies seek to uncover similarities of viewing patterns of large groups of viewers (e.g., over art, or computer-generated scenes), even though variability and task-dependence of eye movements are widely acknowledged. In a nomothetic approach, important concerns are generalizability of results to larger populations and the selection of appropriate population representatives.

A particularly instructive example of a (difficult) field study involving a specific (somewhat idiographic in spirit) population was presented by Hornof and Cavender (2005) who investigated an interactive eye drawing application designed for children with severe motor impairments. In a pilot study, Hornof and Cavender first employed ten participants without disabilities where half were children (average age of 12; the other half's average age was 26). (Prior to this, the authors performed two user observation studies, one with children and adults without disabilities, the other with adults with severe motor disabilities.) Some users were based locally (presumably in the lab), others were at remote locations. The final evaluation study was performed by four participants from the target audience, aged 9, 12, 18, and 61. The difficulty here is of course selection of a representative sample population. Initially, it makes sense to perform prototypical evaluations in a highly controlled environment (lab) with population samples that may not necessarily generalize to the target population (e.g., adults without disabilities). Although both constraints present generalizability problems in terms of environment and individuals' abilities, this operational constraint makes sense: gross problems can be identified early in the prototypical development stages, before moving on to field trials with members of the target audience.

13.2.4 Sample Population Versus Single-Case Experiment Versus Case Study

Selection of a sample population is generally performed with the intention of generalizing results to a wider (if not potentially global) population. There are instances when it is appropriate to consider a population consisting of a single individual. This may be a case study of an individual, as is often reported in clinical accounts. Phineas P. Gage presented the famous case of an individual accidentally lobotomized by an iron rod propelled through his frontal brain regions in a rock blasting accident. What was interesting about this case was the change in Gage's personality following the accident. Prior to this accident Gage was characterized as mild-mannered, however, following the accident he had become aggressive, rude, and "...indulging at times in the grossest profanity (which was not previously his custom), manifesting but little deference for his fellows, impatient of restraint or advice when it conflicts with his desires...," according to the physician Harlow in 1868 (Harlow, 1868). Analogously, there may be highly interesting and informative eye tracking clinical cases.

An alternative to the case study is the single-case experiment. This is a quasi-experimental design in that it lacks randomized allocation of participants to treatment conditions. This design is appropriate for specific individuals or small groups of people, e.g., experts, as they may be employed in heuristic usability evaluations. Performance of the expert(s) can then be compared to the average performance of novices as a type of baseline comparison under similar or identical conditions. This type of approach may be taken in training and assessment studies.

13.2.5 Within-Subjects Versus Between-Subjects

Of the many experimental approaches available, the two most likely methods of collecting data from groups is either via a within-subjects (repeated measures) or via a between-subjects design. A within-subjects design uses one group of participants and tests them under all treatment conditions. A between-subjects design uses different groups of participants, where different treatments are assigned to different groups.

The within-subjects design repeats treatments per individual, hence it is also known as the repeated-measures design. The most prominent problem with repeated measures is that the analysis of results will suffer from order effects unless care is taken to counterbalance the conditions. Order effects may include fatigue or learning, for example. To counterbalance the conditions, a Latin square may be used to (randomly) assign conditions to participants. The Latin square of order n is an $n \times n$ array in which each cell of the array contains one of a set of n symbols such that each symbol occurs only once in each row and column; e.g., given $n = 4$ conditions, A, B, C, D, the following Latin square is generated,

A	B	C	D
B	C	D	A
C	D	A	B
D	A	B	C

where each row can be used to assign a treatment sequence to each of four study participants.

A between-subjects design can avoid the repetition of treatments by using different groups of participants per treatment. For a four-condition experiment, four groups could be used, with different people assigned to each group. Care has to be taken to avoid accidental homogeneity of groups, e.g., testing two groups where one group is entirely male, the other entirely female, introduces gender bias into the results. Random assignment may not attenuate participant variability fully, however. Other strategies for group assignment involve prescreening of participants, and/or targeted assignment by representation. The former involves some form of participant assessment, e.g., questionnaire or pretest. The latter may involve assignment ensuring roughly equal representation of disparate individuals in each group (e.g., equal representation of people with 20/20 vision and myopes in each group).

There are two disadvantages to the between-subjects design. First, more subjects are needed to obtain similar power to a within-subjects design. Recruiting and running more subjects can be costly and time consuming. Second, if there is too much variance among the participant groups, statistical analysis may be complicated; e.g., it may not be possible to perform parametric evaluation of the statistical mean differences.

13.2.6 Example Designs

A general discussion of experimental designs is complicated without a specific context and specification of IVs/DVs being tested. It is often easier to settle on the IVs/DVs first and then develop the design. Doing so will stipulate the requirements for the number of groups or experimental trials required, depending generally on whether a within- or between-subjects design is adopted. In this section some basic designs are offered. Not all are applicable to eye tracking studies, but they are given for a better sense of completeness.

Single Individual, Time Series

This design is exemplified by the traditional drug effectiveness experiment. In this design, there is just one participant, and measurements are taken over time. Prior to administration of a treatment, a baseline measurement is taken. Subsequent measurements are then compared to the baseline to test for the drug's effect. Figure 13.1 is an example of a simple *ABAB* type design where *A* indicates no treatment (baseline) and *B* indicates treatment. In Figure 13.1 only *ABA* is shown. The dashed lines indicate administration and subsequent cessation of treatment. The graph itself would suggest some level of the dependent variable being measured (perhaps some form of subjective well-being or alternatively some physically measurable indicator such as blood pressure). Analysis of this type of experiment typically requires comparison

Fig. 13.1. Example of single-subject, time series *ABAB* type design.

of the DV mean during the specific time sequences during which it was known the drug was taken, e.g., time period over which treatment B was active. If this mean is statistically different from the mean during the period when the drug is not present, i.e., period of treatment A, then the drug is said to have an effect.

Note that this design can be thought of as either one with two independent variables, A and B, or just one but with two levels. Relabeling the diagram in Figure 13.1 with A_o and A_1 would indicate one treatment, or *factor*, administered at two levels: A_o meaning absent, A_1 meaning present. This would now be considered a single factor design.

General Pre–Post Design

The single individual design can be adapted to a single group design, wherein the same group of individuals is given the treatment in the AB treatment administration sequence, and the data are now analyzed over the entire sample of participants. This design is known as pre–post because measurements are taken prior to and following treatment administration. This design is shown in block form, along with other designs, in Figure 13.2. Once again, this too can be considered a single factor design.

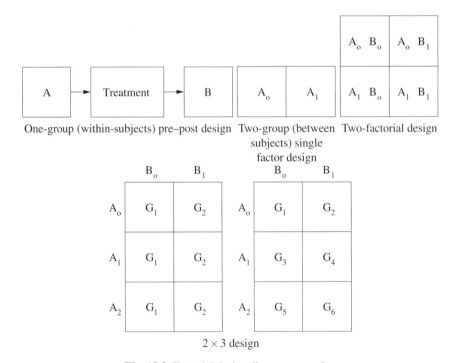

Fig. 13.2. Factorial design diagram examples.

Two-Group Design

The previous single factor (one experimental variable) designs are within-subjects because all participants (including the individual, naturally) are given the treatment. Extending those designs to between-subjects leads to a two-group factorial design, shown in block form in Figure 13.2. Each group, identified by the boxes, receives either A_o or A_1, where A_o is a placebo and A_1 is the treatment. Being a between-subjects design, care must be taken in assignment of participants to groups. This is accomplished via randomization, prescreening, or targeted assignment, as discussed above. Comparison of the mean response to the stimulus (A_1) versus the mean response to the placebo (A_o) would determine significance (or insignificance) of the treatment (i.e., its effect).

Two-Factorial

Consider the four boxed two-factorial design in Figure 13.2. There are four experimental conditions, composed of a combination of two experimental variables, each administered at one of two levels. This is a 2×2 factorial design, where the 2 indicates the number of levels of each factor. What is important is how groups of participants are allocated to each condition. Several permutations are possible:

- If each box in the 2×2 cube is assigned a unique group of participants, this design is a 2×2 four-group, or fully independent, between-subjects (or just 2×2 between-subjects) design.
- If we assigned one group of participants to condition B_o (the left column) and another group to condition B_1 (right column), then the design becomes a 2×2 mixed design (because it mixes within- and between-subjects). Because the same group would receive treatment B_o or B_1, repeated measures analysis would be required to test for significance between these factorial levels inasmuch as the measurements of B are no longer independent.

2×3 Factorial Design

Finally, consider the 2×3 design shown at the bottom of Figure 13.2 in yet another form of representation, where groups of participants are now made explicit with the subscripted G notation. Both are 2×3 designs because there are two levels of B and three levels of A. The design at left, however, differs because treatment A varies within-subjects. Analysis of A's effect, therefore, examined across rows, will require repeated measures, or nonindependent analysis across the rows of the data collected. Analysis of B's effect, examined across columns, however, can be performed via independent analysis methods because this treatment is administered between-subjects. In contrast, the 2×3 design at right is fully independent, or fully between-subjects.

13.3 Measurement and Analysis

Experimental design is generally concerned with the manipulation of independent variables and the subsequent measurement of dependent variables. The general expectation (hypothesis) is, of course, that manipulation of the IV will render some effect on the DV and that, most importantly, the effect will be measurable and (statistically) significant. The "trick" to conducting a meaningful experiment is the operational definition of the dependent variable, that is, that some meaningful measurement of the expected effect can be defined.

To give an example related to eye tracking, we could test the attentional quality of a banner advertisement on a Web page. To do so, we hypothesize that a dynamic blinking banner ad is more visually attractive than a static image. This hypothesis is based on the underlying theory of low-level vision which tells us that vision, particularly in the periphery, is sensitive to "sudden onset" stimuli. Thus we have the expected cause and effect that we operationalize by first defining the IV to be the static or dynamic nature of banner ad, i.e., presence of motion in the stimulus (one can be much more specific and stipulate such details as the frequency of the animation, its size, position on the Web page, and so on; this would lead to an increase in the number of IVs being manipulated).

The definition of the DV should match our initial qualitative description of "attentional quality". We can do so by quantifying visual attention in terms of the number of fixations devoted to the banner ad during a given interaction (Web surfing) session. Note that this operationalization of the DV is quite specific in its assumption of fixations denoting visual attention. A common criticism of this assumption is that it is possible to voluntarily disengage attention from foveal vision. This is true, but because we cannot measure this covert mechanism of attention, the best we can do is acknowledge awareness of this fact and stipulate that we are only measuring overt attention and assuming that during the course of our study we expect participants' attention to be aligned with foveal vision.

Other important details must also be addressed. In particular, the task that participants are going to perform needs to be defined. For example, navigation or search may affect the Web task and hence recorded eye movements differently. Additionally, remaining procedural details must also be specified; e.g., how long should individual trials last, how many trials should each individual perform, etc. (These will to a certain extent depend on the complexity of the design in terms of number of IVs.) Specific to eye tracking, calibration is important: how many points are to be used, and how often is calibration to be performed?

The test of a hypothesis depends on a comparison of the outcomes of the dependent variables, following manipulation of the IVs. In the banner ad example, the DV was operationalized quantitatively as a number of fixations. Given the two conditions, the static and dynamic ad, we would obtain two groups of measurements: number of fixations per participant group per condition. Because there are only two outcomes, and we have numerical data, we can obtain descriptive statistics about the two results and perform parametric tests on the outcome differences. For example, as is fairly common, means and variances would be computed for both groups of fixations. Subsequently, a t-test could be made on the pair of means to test for statistical significance in the difference of means.

The t-test basically reports whether the two sets of numbers overlap in a statistical sense, based on the assumption of normality of the data. That is, the data are assumed to fit a normal (Gaussian) distribution. The t-test checks whether the two distributions overlap. If they do not, one can then claim, at a sufficient level of significance as reported by the t-test, that the outcomes differ, and therefore, an effect is observed. In the hypothetical banner ad example, if it turned out that the number of fixations falling on the dynamic ad was significantly greater than the number of fixations falling on the static ad, then we could state that our data support the initial hypothesis that the dynamic blinking ad is more visually attractive by drawing to it a significantly higher number of fixations. (More formally, the null hypothesis H_0 would have been given as one stating no expected difference between outcomes, which would have to be rejected in favor of our stated alternative hypothesis that the outcomes do differ.)

Given more than two experimental conditions, e.g., dynamic ad, static ad, no ad (a control condition), with all other experimental procedures remaining unchanged, the statistical parametric test that is used quite often as a test for significance is the Analysis Of Variance, or ANOVA. ANOVA is conceptually an extension of the t-test in that it only tests for overlap of the data distributions (assumed to be normal). Its main function is to report whether there is significant distance between overlap of the means. For example, given our three ad conditions, let's say we collected five fixations per group (and recorded them as either within the banner ad or not). We would then plug the resultant data ($n = 15$) from the three groups into a statistical program such as SPSS, S-Plus, or R and obtain an ANOVA table. To find out whether there is sufficient dispersion in the means, i.e., a significant difference between the means, the F statistic and level of significance are examined. In this example, suppose $F = 4.761$ at significance level 0.030. This result is significant, and would be reported as $F(2, 12) = 4.761, p < 0.05$, where the numbers in parentheses following F indicate the between- and within-group degrees of freedom (df), respectively, whose total should add up to $n - 1$. In this case there were three groups and $n = 15$, giving between-groups df $= 2$ and within-groups df $= 12$ so that $2 + 12 = n - 1$.

ANOVA is a particularly popular analysis tool and is used as a preliminary indicator of (statistically significant) effect. However, it only reports a difference among the means. To pin down which mean (and hence condition) differs from all others (perhaps all conditions differ significantly from each other), usually a pairwise comparison of means is then required, such as the pairwise t-test or the pairwise Kruskal–Wallis test if the data cannot be assumed to be normally distributed (or the measurement scale of the data is not an equal interval scale or the sampled data do not have approximately equal variances).

In general, the type of statistical test of difference performed depends on the type of data being collected and the number of samples (groups) measured. Table 13.1 lists statistical tests of difference for sample pairs. Table 13.2 lists statistical tests of difference for multivariate data, in this context meaning two or more variable quantities. Nominal data are usually not measured, and refer to sorting or assignment of

Table 13.1. Statistical tests of difference of sample pairs (df = 1).

	Samples	
Measurement Level	**Independent**	**NonIndependent**
Nominal	Chi-square	(Binomial) Sign test
Ordinal	Mann–Whitney U test	Wilcoxon signed ranks
Parametric	z-test, t-test for independent means	t-test

Table 13.2. Statistical tests of difference of multivariate data (df > 1).

	Samples	
Measurement Level	**Independent**	**NonIndependent**
Nonparametric	Chi-square	Kruskal–Wallis test
Parametric	ANOVA	ANOVA

values into categories, hence they are also known as categorical data (e.g., number of blue-eyed people in the sample population). Ordinal data are measured but only denote the order or position of a data point (e.g., the top five scorers on a midterm exam).

Generally speaking, eye movement data are considered parametric because related metrics can be represented by a uniform (equal distance) interval/ratio scale. An interval scale is one composed of equal units, where, for example, the distance from 160–165 cm is the same (*means* the same) as that between 170–175 cm. Related to the interval scale is the ratio scale, where the latter is an interval scale with a necessary and absolute zero. Examples of ratio scales include reaction times or distance. Timing starts from zero, and in this scale it makes sense to say that if something completes in half the time of something else, then it is twice as fast. Note that in the above eye movement example, the number of fixations can only be considered on this scale if it makes sense to say that twice as many fixations counted on one ad is in some sense twice as meaningful or valuable as on the other (and zero fixations is also meaningful). Fixations can be interpreted in this way if we consider them as indicators of cognitive load; e.g., devoting twice as many fixations to a region may mean that twice as much cognitive effort is being exerted (an operational assumption). This might not always be a valid interpretation, however. For example, twice as many fixations on some conspicuous portion of the screen (e.g., an ad) may mean the viewer is bored or distracted by something entirely different from the ad content, and hence is not indicative of cognitive load at all (just the opposite!). Thus one must be very careful in considering the type of measurement being recorded, and how the variable is being operationalized.

13.4 Summary and Further Reading

This chapter attempted to provide a rather general overview of experimental design. Being a very broad topic in itself, the resulting summary is most certainly incomplete. However, the main points that appear to be relevant to eye tracking are as follows.

1. Start any experiment with a good hypothesis statement; this is a good strategy for most experimental designs, not only ones concerned with eye tracking. For eye movement work specifically, think in terms of the kind of data that the tracker is able to provide, e.g., fixations, fixation durations, etc.

2. This summary tended to favor experiments over observational studies, however, eye tracking observational studies are not ruled out by any means. It is quite conceivable that observational studies can be performed to formulate an initial impression of how people may look at some visual stimulus. For example, eye movements collected over a prototype of an interface may guide the designer on the interface's layout. Similarly, layouts of Web pages, advertisements, or other visually rich media may be evaluated in this manner.

3. Due to the nature of the equipment, most eye tracking studies are lab-based, although due to the emergence of cheap head- or body-worn equipment, one can conduct eye tracking studies in the field. For a fascinating example of field studies (literally!), see the eye tracking work with lemurs by Shepherd et al. (2004). (It seems lemurs tend to look at the tails and heads of other lemurs.)

4. The issue of idiographic versus nomothetic study, similar to sample population versus single-case versus case study design depends on the situation (as most experimental design considerations do anyway). The point is, it would certainly be desirable to generalize eye movement behavior to large populations, e.g., all Web users. However, in some cases, this may not make sense. As in Hornof and Cavender's (2005) studies, the target population was quite specific and therefore a restricted sample population made sense. A good guideline here would be to consider who the target audience is in selecting the sample population during the design of the experiment.

5. The experimental design review tends to favor factorial designs. Indeed, a good deal of eye tracking research appears to employ these designs. Whether they are conducted within- or between-subjects depends on the operationalization of the independent and dependent variables and other constraints of the experiment (e.g., time and money being the notorious ones).

6. The review of data analysis techniques attempted to be as complete as the competing goal of being succinct would allow. The general (and potentially dangerous) recommendation is the use of ANOVA as the main test of multivariate statistical difference. If ANOVA reports statistical significance, ensuing pairwise t-tests or Kruskal–Wallis tests should follow. The potential danger with this advice is that ANOVA may then be seen as a hammer with all experimental data nails. Although ANOVA is popular in the eye movement literature, the most conservative advice that can be given is to use the right tool for the job: carefully select the appropriate statistical analysis tool for the data collected.

The reader is strongly encouraged to delve into the rich literature on experimental design and/or attend experimental psychology lectures, if at all possible (inspiration for this chapter grew from personal notes taken during two guest lectures given by psychologist Eugene H. Galluscio in my eye tracking methodology course at Clemson University, Fall 1999). A good place to start is Coolican's (1996) introductory text.

Beyond experimental psychology references, especially for human–computer interaction work, where eye tracking is by now fairly commonplace, consider Stone et al.'s (2005) book on User Interface (UI) evaluation or Dix et al.'s (2004) text on Human–Computer Interaction (HCI).

14

Suggested Empirical Guidelines

This chapter attempts to provide guidelines for conducting an eye tracking experiment, e.g., performing a usability study incorporating eye movements. It is not possible to provide a cookbook recipe for doing so, however. The experiment must be designed within the context of the phenomenon under investigation. Nevertheless, a general outline can be provided to help guide the process of experimental design. Here, the discussion is biased somewhat towards *performance measurement*, a specific form of testing method often seen in human factors research, particularly as practiced within computer science, and more specifically still within the human–computer interaction community. Other testing methods not addressed here include *ethnographic studies*, *co-discovery*, or *think-aloud*.

Although think-aloud is omitted, it deserves special consideration due to some concerns raised recently about this protocol, particularly in conjunction with eye tracking. Although think-aloud is certainly a valuable usability engineering method, it may change how a user attends to a given task, and more importantly may alter her or his gaze patterns (Bojko, 2005). Before considering concurrent think-aloud, one should acknowledge that the act of thinking aloud is likely to alter gaze patterns as well as performance statistics (because it may take longer to complete the given task while verbalizing). A particularly interesting manner of using think-aloud was proposed by Guan et al. (2006) wherein they suggest that eye movements may be particularly illustrative to the user while verbalizing actions during a retrospective think-aloud session. In this instance, because the task has already been completed, performance metrics are still reliable. More importantly, however, retrospective examination of one's scanpath may trigger recollection of some difficulty the user may have experienced during the session. For example, even though a user may have exhibited recurrent eye movement switching between multiple information points, their verbal report may reduce remembered ocular behavior to a single observation. Replay of their eye movements may serve as a catalyst toward a richer explanation.

The testing methods identified so far form but a subset of methods within the usability measurement framework, depicted in Figure 14.1. The framework includes

Fig. 14.1. Usability measurement framework.

inspection methods (consisting of heuristic evaluation, cognitive walkthrough, pluralistic walkthrough, feature inspection, standards inspection, or guidelines checklist) and inquiry methods (based on *questionnaire and interview protocols*). Shumberger et al. (2005) provide more details on this particular view of usability engineering (see also Mayhew (1999)).

The evaluation plan presented here follows what currently appears to be common practice followed by usability practitioners, as outlined by Stone et al. (2005).

14.1 Evaluation Plan

In general, eye tracking data complement the set of usability metrics common in human factors research, mnemonically easy to remember as the SEE metrics: (1) Satisfaction, (2) Efficiency, (3) Effectiveness. Satisfaction quantifies subjective users' impressions dealing with such indicators as operability and learnability of a given task. Efficiency and effectiveness metrics are objective performance measures of speed and accuracy, respectively. Eye movement metrics, in contrast, are process measures. The number of fixations, fixation durations, and gaze switching behaviors, provide insights into the cognitive state of the user during evaluation tasks. Note that scanpaths intrinsically collect some performance measures by virtue of encoding speed (time to task completion obtained as a difference of the timestamp of last fixation minus that of the first) and accuracy (number of targets fixated).

14.1.1 Data Collection

The following (usability) metrics can be collected during system evaluation.

1. Subjective metrics

a) *Satisfaction.* Following the Software Usability Measurement Inventory (SUMI), questions to users can initially be posed regarding the system's usability:

    i. *Learnability.* "I can learn all the features of this system."

    ii. *Helpfulness.* "The instructions and prompts are helpful."

    iii. *Control.* "I sometimes don't know what to do next with this system."

    iv. *Efficiency.* "If the system stops, it is not easy to get back to what I was doing."

    v. *Affect.* "I would recommend this system to a colleague."

Responses to such questions are usually quantified on a suitable Likert-like (e.g., five-point) scale. The questions themselves need to be tailored to the tasks identified in the given study. For example, questions can address system usability. Other attributes that can be probed include Compliance (to standards, or perceived standards), Operability, Understandability, Learnability, and Attractiveness, easily remembered by the COULA acronym (Shumberger et al., 2005).

2. Objective metrics

  a) *Performance measures.*

    i. *Efficiency.* Time to complete task.

    ii. *Effectiveness.* Number of errors committed.

  b) *Process measures.*

    i. *Number of fixations.* This is an indicator of targets fixated during scene exposure.

    ii. *Fixation durations.* This is an indicator correlated with cognitive function (longer durations indicate an increase in cognitive function).

    iii. *Attentional switching.* This is an indicator of visual attention distribution (e.g., degree of focus or inattention).

    iv. *Scanpath similarity.* This is (potentially) an indicator of position and sequence similarity among different viewers, e.g., suitable especially for expert and novice comparisons. Open-source scanpath comparison tools based on string-editing are just now emerging (West et al., 2006; Heminghous & Duchowski, 2006).

According to Jacob and Karn (2003), the most common set of eye tracking metrics used thus far is listed in Figure 14.2, where AOI refers to Area Of Interest, such as a rectangular ad region on a Web page.

• Fixation	• Scanpath (fixation sequence)
• Fixation (gaze) duration	• Area of interest, or AOI
• Fixation rate (overall)	• Gaze % (prop. of time) per AOI
• Fixation duration mean (overall)	• Number of fixations per AOI
• Number of fixations (overall)	• Gaze duration mean per AOI

Fig. 14.2. Traditional eye tracking metrics.

The traditional metrics given in Figure 14.2 are effectively low-level measures. Although valuable, when given outside an application context (e.g., usability described by ISO 9241), they may fail to impart a sense of meaning to experimental design. That is, how are fixations supposed to augment more commonplace measures of satisfaction, efficiency, and effectiveness? In practice, by providing process metrics, eye movements can help corroborate the performance metrics of efficiency and effectiveness (see Chapters 15 and 18 for examples). Process and performance metrics are generally quantitative. Qualitative evaluation, of say users' satisfaction, can be recorded by questionnaires of users' attitudes toward whatever was being tested during the experiment.

Scanpath similarity metrics, as proposed by Privitera and Stark (2000), can be sorted and stored in a table, named the Y-matrix, having as many rows and columns as the number of different scanpath sequence fixations to be considered. Scanpath comparison values from the Y-matrix (which typically contains large amounts of data) are condensed (averaged) and reported in two tables, called parsing diagrams, one for each of position similarity (S_p) and sequence similarity (S_s) indices. The first table reports S_p statistics, that is, the correlation between two string sequences in terms of attentional loci (the scanpath sequence, i.e., the order of fixations, is itself not considered in this measure). S_p values are generally expected to be higher than the measure reporting on order of fixations, the S_s value. Examples of a Y-matrix and parsing diagrams are shown in Figure 14.3.

S_p	Subj. 1		Subj. 2			S_s	Subj. 1		Subj. 2	
	Pict1	Pict2	Pict 1	Pict 2			Pict1	Pict2	Pict 1	Pict 2
S1P1	R	I	L	G		S1P1	R	I	L	G
S1P2		R	G	L		S1P2		R	G	L
S2P1			R	I		S2P1			R	I
S2P2				R		S2P2				R

Same Subj.	Diff. Subj.		Same Subj.	Diff. Subj.
Repetitive	Local	← Same Image (SI) →	Repetitive	Local
Idiosyncratic	Global	← Diff. Image (DI) →	Idiosyncratic	Global
S_p	Random		S_s	Random

Fig. 14.3. Scanpath comparison Y-matrices and parsing diagrams. See Privitera and Stark (2000) for details.

Each of the parsing diagrams reports several correlation measures: **Repetitive, Idiosyncratic, Local,** and **Global.** Repetitive values repeat an individual's propensity to view a specific image in the same way. Idiosyncratic values report on the within-subject attentional scanning tendencies of individual subjects; i.e., these values report correlation between scanpaths made over different pictures by the same subject. For example, in reading studies, English readers would be expected to exhibit high idiosyncratic indices due to the adopted left-to-right text scanning strategies. Local indices report on between-subject correlations of scanning patterns over similar stimuli. That is, local indices report on different subjects' scanpaths over the same picture.

When comparing experts to novices, local indices are the most important because these indices divulge the correlation between expert and novice scanpaths made over the same images.

Global measures report on the correlation between scanpaths made by different subjects over different stimuli. Should these values be highly correlated, this would suggest that stimulus images tend to be viewed similarly by different people. This indicator may be highly relevant for empirical evaluation of synthetic (computer-generated) imagery. For example, in the quest for photorealism, identification of image attributes contributing toward the perception of realism may be quite valuable.

14.1.2 System Identification

The system to be used for eye movement capture needs to be identified and reported. This is usually listed in the "apparatus" section of a study report and should include the operating characteristics of the eye tracker (see Chapter 13).

Note that apart from the eye tracking apparatus used to obtain eye movement recordings, one must also identify and report the stimulus with which the participants are working. This will probably be related to the main point of the study and may be elaborated upon elsewhere in the evaluation plan and subsequent report, however, it must be done in sufficient detail to allow reproducibility. For example, if conducting a study on the photorealistic qualities of some computer-generated image, the techniques used to generate the images must be described in sufficient detail to allow reproduction and similarly for other stimuli such as Web page banner ads (what were the ads, where were they located, etc.), advertising copy, and so on.

14.1.3 Constraints

There are (at least) three operational constraints associated with system evaluation: time, personnel, and money. First, the project is usually subject to a linear timeline: program development precedes evaluation, or in an iterative sense, some prototype must be developed prior to evaluation. Time must be budgeted to allow for both development and evaluation. Generally speaking, plan on evaluation consuming an equal if not greater amount of time than development. Time is needed for running participants through the trials as well as for data analysis. Both aspects are time consuming.

Second, personnel pose the next constraint. Programmers must be available to develop the program/system/prototypes for evaluation. Second, participants must be selected/recruited for evaluation. In academic settings, development is usually left to those knowledgeable with the system requirements, e.g., programmers for development of a software package. For evaluation, student participants are readily available, however, their results may not generalize to the target audience (e.g., the disabled, or personnel with specific expertise such as law enforcement personnel, aircraft inspection personnel, etc.).

A good strategy for research group formulation is an interdisciplinary one. That is, when forming research groups to conduct an eye tracking experiment (or taking part in an eye tracking college course), it is a good idea to form a team such that its members contribute the necessary expertise to see the study through to completion. For example, at minimum the team should include someone knowledgeable in experimental design and analysis to guide the experiment and should also include someone capable of developing the technical environment for testing (i.e., software environment).

Third, money, or rather its lack, is an obvious obstacle. It is needed for equipment and hiring of developers, as well as payment of participants. If conducting expert/novice comparisons, domain experts should be identified and recruited for data collection (and paid if need be). To record data from experts (or at least domain knowledgeable participants), consider collecting data on-site by bringing a portable eye tracker and laptop (e.g., the Tobii eye tracker can be packed in a flight security case for fairly easy transport).

14.1.4 User Selection

User selection poses somewhat of an evaluation constraint, as noted above. Although perhaps readily available, college students may not be domain-knowledgeable. Beyond expertise considerations, one must balance the often competing goals of maximizing the number of participants for eventual statistical power during analysis and minimizing the number of participants for study tractability. Clearly planning for a large number of participants will increase the expected duration of the study. User selection therefore is related to study duration, which is one of the three important operational constraints.

14.1.5 Evaluation Locale

The evaluation locale must be decided prior to conducting the study. It may seem obvious that if an eye tracking laboratory is available, such as what is available at Clemson University shown in Figure 14.4, then that is where the study will need to take place. However, because this is an instructional as well as a research lab, care must be taken to prevent scheduling conflicts and therefore, in a class setting, a sign-up sheet should be used.

Alternatively, it may be possible to either use a head-mounted eye tracker or port a table-mounted eye tracker on-site to conduct an in-field eye tracking study.

Fig. 14.4. Eye Tracking Lab at Clemson University.

14.1.6 Task Selection

Task selection may be the most critical consideration of all, particularly for eye tracking studies. Because eye movements (and attention) are deployed as a combination of bottom-up (stimulus-driven) and top-down (goal-driven) cognitive processes, the nature of the task will influence eye tracking outcomes (the scanpaths). Eye movements are task-dependent and therefore the tasks must be chosen with care.

Task selection depends on whatever human activity is being studied. Eye movements will generally be related to whatever action the participant is involved in, and therefore, the visual task may be rather specific. For example, visual search is a very specific visual task, and is distinct from other tasks such as "free viewing" a piece of art.

Because the nature of the task will influence recorded eye movements, the task(s) issued to participants must be clearly defined and subsequently stipulated in the research report. During the actual running of the experiment, to maintain consistency, it is often a good idea to script the task instructions to participants.

Related to the tasks given to participants, the type of stimulus presented to participants must also be decided ahead of time and reported in sufficient detail following the study. Stimulus types include static images (how many, how were they created, what are the differences between them, etc.), dynamic images (image sequences, videos), Web pages (browser), virtual environments, natural environments, etc.

How should tasks be selected, defined, and scripted for participants? In general, this will depend on the nature of the activity being evaluated. There are tools and strategies that can help in this regard, falling under the objective of task analysis. Although this is a rich topic in itself, Dix et al. (2004) provide the following succinct overview of task analysis techniques and sources of information.

- Task analysis techniques:
    - Decomposition of tasks into subtasks

- Taxonomic classification of task knowledge
- Listing actions performed and tools used
- Sources of information:
  - Existing documentation
  - Observation
  - Interviews

One approach for task analysis is to decompose high-level tasks to a set of low-level ones. Hierarchical Task Analysis, or HTA, is a formal approach for doing so. For example, the task of making a peanut butter and jelly sandwich may involve substeps of fetching the bread, peanut butter, and jelly. Then (optionally) toasting the bread, spreading the peanut butter, jelly, and so on. Obtaining the hierarchy for a given task may involve observing someone else perform it (as the source of information), or perhaps reading an appropriate manual.

14.2 Practical Advice

When it comes to collecting eye movement data, it is a good idea to follow the KISS principle: Keep It Short and Simple. Although a good rule of thumb in general, in the context of eye tracking, KISS is particularly important for reasons of data manageability. Specifically, KISS suggests that eye tracking tasks be limited in duration to minutes if not seconds, rather than hours. Thus, KISS advocates a divide-and-conquer approach to task definition.

There are practical reasons for keeping experimental trials short. First, as sampling rates increase so does the volume of recorded data. Even at a modest 50 Hz sampling rate, data accumulate quickly, not only in terms of eye movements, but also in terms of screenshots, videos, and other related artifacts. Compounding this over multiple participants increases demand for computing resources.

Second, keeping experimental trials short allows frequent calibration. Calibration is still a necessary evil, and it is the only way to ensure the highest accuracy of recorded data. Follow American eye tracking pioneer Lawrence Stark's advice: "Calibrate, calibrate, calibrate." This is sage advice even if the eye tracker can store calibration data (it may be difficult to ensure that a returning participant sits in a similar position to that of their stored calibration).

Generally, one can break down a long user task into shorter subtasks. This will facilitate data management and analysis. Task analysis techniques (e.g., Hierarchical Task Analysis, or HTA), can be used for this purpose. Decomposition of long tasks into shorter ones may yield interesting results that may have been missed if the task had not been decomposed. Chapter 15 reviews a short usability study that relied on task analysis of a relatively long pilot study. It is quite possible that interesting observations resulting from the study analysis would not have emerged had the original pilot study not been reduced in the spirit of the KISS principle.

However, advocacy for the KISS methodology must be tempered by the realization that artificially shortening task duration, or just simply limiting task duration, may adversely affect natural behavior. That is, a similar criticism can be levied against KISS as against think-aloud in the sense that behavioral realism or fidelity may be degraded through artificial alteration of a given task, e.g., Web browsing.

14.3 Considering Dynamic Stimulus

Commercially available analysis tools are beginning to offer means for analysis of eye movements over dynamic stimuli. At the moment, this includes some dynamic Web content, sequences of images, and video clips. Of course, a wide array of dynamic stimuli is gaining interest, including driving and flight simulators, games, office productivity software, mobile devices, VR, and TV/DVD applications, to name a few.

The difficulty in analyzing eye movements captured while viewing and interacting with dynamic stimuli lies in synchronization between real-time eye movements and the dynamic events that took place at the instance of capture. Although commercially available tools handle some aspects of dynamic stimulus generation, for other situations companies may provide a Software Development Kit, or SDK, leaving the development responsibility to the evaluator. Therefore, the onus is on the experimenters to reinstrument whatever applications on whcih they wish to conduct eye tracking studies.

Although reinstrumentation of a given application is undoubtedly the most complex and labor-intensive endeavor, it is also one that affords the most flexibility in data capture and analysis. Chapter 15 provides results from an example of an eye tracking application (a driving simulator) that was developed from scratch with the purpose of collecting eye movements.

14.4 Summary and Further Reading

There do not seem to be, at the moment, any more specific guidelines for conducting an eye tracking study than what was given in this chapter. This may be in part due to the clash of specificity of experiments and the typical generality of guidelines. That is, it is difficult to provide general guidelines to suit all specific eye tracking experimental designs. Nevertheless, the aim here was to provide something of a workable context or framework for designing an eye tracking study. It should be understood that an eye tracking study is not so much different from any other typical human–computer interaction study, it just happens to use an additional specialized tool to collect data. The eye tracker can at a basic level be considered not dissimilar from a sensor measuring pulse rate or blood pressure. As such, the eye tracker can probably complement a large number of evaluation techniques, not just performance measurement. Most up-to-date HCI textbooks that include a discussion of evaluation techniques should provide various alternatives to performance measurement. Dix et al.'s (2004) recent HCI textbook devotes a chapter to evaluation techniques.

Another source of eye tracking study examples is the eye tracking research literature. Since the first edition of this book was written, a good deal of work has appeared, published in conference proceedings such as ACM ETRA and SIGCHI. Numerous eye tracking studies have been published in which stimuli and tasks were combined in interesting ways. The fourth part of this text reviews work from several disciplines. Chapter 15 gives detailed examples of select diagnostic and interactive experiments.

15

Case Studies

This chapter presents summaries of studies conducted with various equipment housed in Clemson's eye tracking laboratory. Study examples include the measurement of eye movements in head-mounted Virtual Reality (VR) and in desktop applications. VR applications used the binocular eye tracker embedded in the helmet-mounted display built by Virtual Research/ISCAN (see Chapter 6). Desktop applications used the table-mounted binocular eye tracker from Tobii (see Chapter 9). Typical use of these devices is shown in Figure 15.1.

Fig. 15.1. Head-mounted and desktop (binocular) eye trackers.

All but one of the examples presented are diagnostic in nature inasmuch as the display does not change or otherwise respond to instantaneous gaze information. The first three examples describe immersive (head-mounted) and desktop virtual environments that were developed from scratch and instrumented from the outset to record eye movements. The fourth example uses the desktop eye tracker and its associated commercial off-the-shelf software (ClearView) to evaluate previously developed applications that were not reinstrumented for eye movement collection in any way. The final example is an interactive, or gaze-contingent, application written specifically to evaluate gaze for interactive object selection.

15.1 Head-Mounted VR Diagnostics: Visual Inspection

Sadasivan et al. (2005) used eye movements as a training aid in a visual inspection task in virtual reality. In this study eye movements were also used to corroborate, or rather help explain, performance metrics gathered in the course of training evaluation. As seen in Figure 15.2, an expert inspector's eye movements were recorded in the virtual environment (left) and then stylized for clarity (right). The stylized display depicted the direction of visual search and the amount of time spent at each Region Of Interest (ROI).

Fig. 15.2. Eye tracking data of expert inspector (left), feedforward training display (right). From Sadasivan et al. (2005) © 2005 ACM, Inc. Reprinted by permission.

The hypothesis put forth in the study was that participants receiving scanpath-based, feedforward search training would exhibit better performance, as measured by traditional (speed, accuracy) metrics as well as process measures derived from eye tracking.

The experimental design used in this study was a specialized form of the pretest–posttest control group design with one independent variable, the treatment condition at two levels: training and no training.

Performance and process measures were reported in terms of number of defects (targets) detected (accuracy, in terms of true positives), time to task completion (speed), and eye movements (number of fixations, fixation clusters (groups), and mean fixation duration). As indicated in Figure 15.3 and evaluated by the t-test, the effect of training appears to cause significant improvement in accuracy ($p < 0.05$) with a simultaneous deterioration in time to completion ($p < 0.01$). Eye movements helped explain this classic speed–accuracy tradeoff. Although there was no observed significant difference in the number of fixations or fixation groups, the t-test of the relative difference in the mean fixation duration was significant ($p < 0.05$). These results suggested that participants who received feedforward training significantly increased their mean fixation duration resulting in a slower-paced inspection strategy.

(a) True positives.   (b) Time to complete.   (c) Fixation points.   (d) Fixation groups.   (e) Fixation duration.

Fig. 15.3. Performance and process measures: relative difference (%). From Sadasivan et al. (2005) © 2005 ACM, Inc. Reprinted by permission.

The given interpretation of these results was that novices learned a more deliberate target search/discrimination strategy that required more time to execute.

15.1.1 Case Study Notes

There are several limitations of this study, particularly in terms of generalizability. First, because the experiment involved college students, it is not known whether results extend to professionals in the field. It is reasonable to expect that they do because the training examined here is essentially by example which is expected to transfer to the actual task.

Second, the more difficult question of training transfer can be directed toward virtual reality itself: does training in VR readily transfer to the physical realm? The results of the study must inherently be restricted to reporting the effects of training in virtual reality. To examine whether training transfers between the simulated and actual tasks can be posed to most simulators where training transfer is assumed or has otherwise been validated. For example, flight simulators have achieved sufficient sophistication so that they are in some cases used for assessment as well as training. The visual inspection environment used by Sadasivan et al. had been previously evaluated for evoking feelings of "presence" (e.g., "being there") suggesting at least subjective fidelity to the actual task. However, one would need to conduct performance evaluations in the field to completely address this issue.

15.2 Head-Mounted VR Diagnostics: 3D Maze Navigation

The diagnostic study in VR performed by Vembar et al. (2004) used eye movements to test users' utilization of an exocentric overhead 2D map for navigation within a virtual maze environment. The map was a type of "you are here" map wherein the independent variable manipulated was the presence of a directional cue (arrow). The arrow within the maze was either absent (control condition), directionally ambiguous (a dot instead of an arrow), or the arrow indicating the viewer's orientation in the maze (one factor at three levels).

The task given to participants was the same: navigate as quickly as possible toward the center of the maze. Based on previous wayfinding literature, Vembar et al. hypothesized that the directional cue (arrow) would afford the fastest navigation performance, as measured by time to task completion.

Of 24 participants originally recruited, data from 9 were rejected due to problematic recording (verified by calculating aggregate fixation error over calibration targets). The remaining 15 subjects were randomly assigned to one of three groups, promoting between-subjects analysis.

The test environment, sample recorded and analyzed gaze data, and performance metrics are shown in Figure 15.4. One-way ANOVA between subjects showed signif-

Fig. 15.4. Simple 3D maze with 2D map, fixations, results. From Vembar et al. (2004) © 2004 Eurographics Association. Reprinted by permission.

icance effect of the navigational cue on mean completion time ($F(2, 14) = 5.29, p < 0.05$). Participants given either the directional arrow or directionally ambiguous cue performed faster than with no cue. Posthoc analysis suggested that gender appeared to influence completion time, with males performing faster than females.

Although gender effects are harder to explain (perhaps gaming or other forms of experience may have played a part), eye movements clearly showed how participants made use of the maze map. Although participants were expected to look at the maze, it was interesting to find that participants given the dot or arrow visual cues would almost exclusively use the 2D map to navigate the maze hardly needing to look at the 3D maze itself. Consequently, it was observed that participants would walk the 3D maze looking at the floor. This blind navigation could be attributed to the overreliance of the participants on the 2D map and the simplicity of the 3D environment.

15.2.1 Case Study Notes

This was a relatively simple but poignant example of using eye movements to gauge the visual accessibility of an informative visual element. Its simplicity (e.g., stark appearance of the maze environment) may be a valid criticism of the study, however, its direct application of eye movements presents a good example of a straightforward, and quick eye tracking experiment. The study, from conception to analysis was completed in roughly one semester (four months).

15.3 Desktop VR Diagnostics: Driving Simulator

Balk et al. (2006) conducted a study to evaluate how eye movements change when talking on a mobile phone while driving. To do so, they created a low-fidelity desktop VR driving simulator, instrumented to collect eye movements during simulation trials. The simulator and collected scanpaths are shown in Figure 15.5. Half of the

Fig. 15.5. Low-fidelity desktop VR driving simulator and scanpaths. Courtesy of Stacy Balk, Kristin Moore, William Spearman, and Jay Steele.

16 university student participants viewed dynamic driving scenes while performing a simulated talking-on-a-mobile-phone-while-driving (TMWD) task. The task simulated TMWD by playing audio excerpts from a Japanese language instruction CD. Participants were asked to follow the language instruction audio by speaking when the audio asked to repeat phrases while simultaneously looking for potentially hazardous events in the scenes (eminent collisions, tailgating). Scenes contained four or seven vehicles (all same model, color), giving a within-subjects 2×2 factorial design with the number of vehicles and TMWD task forming the IVs. All participants viewed 12 trials with four vehicles and 12 trials with seven vehicles and then answered multiple choice questions regarding the driving scenarios. A Tobii ET-1750 eye tracker (50 Hz, .5° accuracy, 1280×1024 display) was used to collect eye movements with five-point calibrations performed at the beginning and in the middle of each session.

Performance measures indicated significant differences between conditions, as indicated by post-driving questionnaire responses. Repeated measures ANOVA revealed a significant effect in the number of vehicles in the scene ($F(1,380) = 11.86, p < 0.01$). Overall, more questions (60.9%) were answered correctly with four-vehicle scenes than with seven-vehicle scenes (44.3%). A significant effect was also seen between the TMWD and driving-only conditions ($F(1,14) = 49.59, p < 0.01$). Participants in the TMWD condition answered fewer postscene questions correctly than people in the driving-only (control) condition (38.5% vs. 66.7%).

In this dynamic environment, a critical component of analysis was registering fixations over moving vehicles. To facilitate this, the simulator was instrumented to count fixations over the vehicles' 150×150 pixel 2D projected bounding boxes (Regions Of Interest, or ROIs). Repeated measures ANOVA showed a significance difference in the percentage of fixations on vehicles involved in hazardous events between driving conditions ($F(1,14) = 7.37, p < 0.05$). Throughout the trial the percentage was greater for the driving-only condition (40% vs. 29%). ANOVA of mean fixation duration also showed that the overall mean duration of fixations in ROIs was greater for the driving-only condition (9.6 s vs. 6.5 s, on average, $F(1,14) = 6.51, p < 0.05$).

Participants in the driving-only condition devoted more fixations and time in the ROIs (both during the hazardous event and the entire trial) than people in the TMWD condition. This is not surprising when taking the number of correct responses for each group into consideration. That is, people in the driving-only group appeared to fixate on the vehicle involved in the critical event more than those in the language group: keeping an eye on the hazardous vehicle, so to speak. This is probably due to one of two reasons: an increased mental workload caused the TMWD group to not follow vehicle location after the event, or they did not see the event occur. This study's findings tended to support Gugerty's (1997) finding that when mental workload is increased, drivers are less able to maintain high situation awareness.

15.3.1 Case Study Notes

This study is a good example of a diagnostic study of a dynamic environment on the desktop (desktop VR). Of particular note is the instrumentation effort required to keep track of users' gaze locations over dynamic stimuli (the moving vehicles). Computationally this can be seen as the registration problem, and in this case is not particularly difficult because it requires calculating the 2D projection of a 3D object's bounding box. Of course development of the 3D gamelike simulator in the first place is nontrivial. The point is that given such an environment (a game engine, basically), it is not particularly difficult to reinstrument it to collect eye movements. One key component that must be included, however, is calibration. This requires providing a "calibrate eye tracker" function which, when invoked, presents a number of calibration points to the user. Modern eye tracking hardware may allow the system to store this calibration which can then be used in subsequent sessions (provided the user sits at about the same location as when calibration was obtained).

15.4 Desktop Diagnostics: Usability

The KISS principle was used in the design of an eye tracking usability study of two related software components for the U.S. Navy.

Apparatus. A Tobii ET-1750 eye tracker was used to collect eye movements during display of the Windows XP desktop using Tobii's ClearView software (Tobii Technology AB, 2003) with "screen" as the stimulus option (at 15 fps). Both the display and eye tracking server ran on a 2.0 GHz Intel Pentium M Dell Inspiron 9300 laptop equipped with 1 GB RAM and an nVidia 6800 Go graphics card.

Subjects. Four U.S. Navy personnel were recruited (3 M, 1F, average age 25). All had 20/20 corrected vision. One was an experienced operator, the rest were novice users. The experiment was conducted on the exhibits floor of the ForceNet conference.

Design. The usability scripts developed for the study were drawn from a mock Software Acceptance Test (SAT) demonstration performed by an expert (software developer). The demonstration lasted approximately 14.5 minutes. At Tobii's 50 Hz sampling rate, this yields about 43,500 raw gaze points. Although analysis (i.e., removal of signal noise, classification of fixations, removal of saccades, and fixation clustering) reduces this figure considerably, the resultant aggregation of fixations may still generate a somewhat blurry picture of attention distribution. For example, Figure 15.6, easily generated by the analysis software, shows the expert's fixation "hotspots" from the entire session over a single representative screen snapshot. Is this type of analysis meaningful, e.g., as informative as Web search's "golden triangle" (Eyetools et al., 2006)? Clearly one can deduce that attention was generally split between the upper-left region (e.g., over menu items) and screen center (e.g., pop-up windows, as shown in this particular freeze frame), but nothing else meaningful readily pops out beyond this simple observation.

To break down the expert's demonstration into manageable chunks for novice users to perform, the task was split into seven subtasks, each renamed as a Software Usability Task, or SUT. The analysis performed to break down the initial ~14.5 min. task resembled Hierarchical Task Analysis (HTA) in spirit but in practice was more informal (formal HTA would probably be quite useful here). The goal was to define tasks that were as atomic as possible, e.g., perform a complete action in each task. On the other hand, the goal was also to keep the subtasks as short as possible. The final seven SUTs partially achieved both goals, although SUTs 001 and 005 were still somewhat long both in duration and number of execution steps. Table 15.1 lists the duration and number of steps of each SUT, as measured for each of the four participants in the study $(S_1 \dots S_4)$ as well as the expert (S_e). SUT-003 and SUT-006 are given in Tables 15.2 and 15.3 in their entirety.

Fig. 15.6. "Hotspots" from expert's ~14.5 min eye tracking session over a single representative screen snapshot.

Table 15.1. Example Software Usability Task durations (in minutes).

Instruction →	Visual		Verbal			
Task ↓ User →	S_1	S_2	S_3	S_4	S_e	# Steps
SUT-001	8.80	8.97	7.40	6.97	4.18	42
SUT-002	2.38	3.22	2.30	2.50	2.05	18
SUT-003	1.48	1.53	1.18	1.25	1.52	4
SUT-004	1.82	1.20	0.88	0.76	0.42	8
SUT-005	6.30	6.30	4.22	4.50	1.60	23
SUT-006	1.42	1.67	0.93	0.73	0.40	7
SUT-007	3.77	3.68	3.05	2.32	2.03	19

Table 15.2. Example Software Usability Test SUT-003: RFI Submission.

User Action	System Response
Step 76: From the main menu, select **DBA** and then choose **RFI Manager**.	**Expected Response:** A new **RFI Manager** window is displayed.
Step 77: From the RFI window's menu, select **File** and then choose **New RFI**.	**Expected Response:** An RFI fields are cleared for a new RFI to be created.
Step 78: Type the following data in the appropriate fields: • Unit – `2mardiv` • RFI Number – `1` • Date Submitted – `181645Z May 00` • LTIOV – `231645Z May 10` • Priority – `1-Flash` (select from drop-down menu) • Status – `OPEN` (select from drop-down menu) • Details – `This is a test of the RFI Submission Process` Select `OK`.	**Expected Response:** The RFI record is added and the scroll buttons become available in the bottom left corner of the window.
Step 78: Close the **RFI Manager** window.	**Expected Response:** The **RFI Manager** window is closed.

The eventual design adopted for this study was a between-subjects factorial design with an independent variable of task instruction (either visual or verbal). Half the participants were assigned randomly to each group.

Procedure. Participants were seated 50 cm in front of the display and were verbally introduced to the nature of the experiment. They were told that speed and accuracy counted, and were instructed to "complete each task as fast as you can with minimal mistakes (e.g., avoid typing errors)." Each user performed all of the seven SUTs. Each trial consisted of prescripted commands that the user needed to execute. None of the steps was designed to generate errors or other artificial problems for the users. All steps generated the expected responses.

The nature of the software being tested necessitated that each participant follow exactly the same order of SUTs (from SUT-001 to SUT-007) because successive steps in the sequence depended on results from previous steps (e.g., entry of database records). Two of the four participants read the instructions and two listened as the instructions were read to them. The expert performed all steps from memory.

Traditional Usability Results. None of the traditional usability metrics is particularly informative. Participants performed all tasks at about the same speed (with the verbally instructed group performing slightly faster because they did not need to devote time to reading). Accuracy metrics also suggest lack of informative value

Table 15.3. Example Software Usability Test SUT-006: NAI Definition.

User Action	System Response
Step 111: Maximize MICA.	**Expected Response:** The MICA Application appears.
Step 112: Expand the **MICA** level. Expand the **United Endeavor** tree item. Expand **Main Attack**. Click on **DIVCP1**.	**Expected Response:** MICA list is expanded. United Endeavor list is expanded. Main Attack is expanded. DIVCP1 is highlighted and the DIVCP1 Details window is displayed in the right pane.
Step 113: Right-click on the first item posted to the PIR group box and select **PIR Details** from the menu.	**Expected Response:** The PIR window opens.
Step 114: In the NAI box on the PIR window, right-click and select **Add NAI**.	**Expected Response:** An Overlay Files and Segments window opens listing Overlay files.
Step 115: Double-click the **MICA** folder in the left pane of the window.	**Expected Response:** The MICA test overlay file appears in the right pane. The file should be named MICA test.mgc.
Step 116: Double-click on **MICA test.mgc** in the right pane to expand. Click on one of the boxes created with C2PC and then select **OK**.	**Expected Response:** An NAI box is added to the NAI section of the PIR window.
Step 117: Close the PIR window.	**Expected Response:** The PIR window is closed.

because none of the participants appeared to make any gross mistakes (e.g., all were adequate typists and sufficiently proficient with the traditional Windows/Icons/Menus/Pointer interface metaphor). A short task-specific questionnaire was administered after each of the seven SUTs and a longer general questionnaire was given following completion of all SUTs. Answers to the questionnaires qualitatively suggest strong satisfaction with the usability of the software (on a five-point Likert scale, users generally either agreed or strongly agreed with questions regarding perceived satisfaction).

In all, eight satisfaction questionnaires were administered to each of the four test participants. Each response sought an agreement opinion from the participant, with the response options given on a five-point Likert scale ranging from "Strongly Disagree" to "Strongly Agree". The first seven sets of four questions were given to participants after completion of each SUT, thus task-specific responses where obtained regarding system usability pertaining to each SUT. The eighth set of 18 questions asked about the user's general impression of the system. Table 15.4 lists the mean task-specific responses over all tasks.

Table 15.4. Task-specific mean responses over all tasks (five-point Likert scale where 1 = strongly disagree, 3 = neutral, and 5 = strongly agree).

	Question	Mean rating (SD)
1,5,9,13,17,21,25	Overall I am satisfied with how easy it was to use this system to accomplish this task.	4.69 (0.47)
2,6,10,14,18,22,26	I was able to complete my task quickly using this system.	4.63 (0.62)
3,7,11,15,19,23,27	I felt comfortable using this system to accomplish this task.	4.69 (0.48)
4,8,12,16,20,24,28	This system had all the functions and capabilities I wanted it to have to accomplish this task.	4.75 (0.45)

Table 15.5. General mean responses (five-point Likert scale where 1 = strongly disagree, 3 = neutral, and 5 = strongly agree).

	Question	Mean rating (SD)
29	Overall, I am satisfied with how easy it is to use this system.	4.25 (0.47)
30	It was simple to use this system.	4.50 (0.58)
31	I can effectively complete my mission using this system.	4.00 (1.15)
32	I am able to complete my mission quickly using this system.	4.00 (1.15)
33	I feel comfortable using this system.	4.50 (0.58)
34	It was easy to learn to use this system.	4.25 (0.50)
35	I believe I became productive quickly using this system.	4.00 (1.15)
36	The gives useful error messages that clearly tell me how to fix problems.	3.00 (0.82)
37	When I make a mistake using this system, I can recover easily and quickly.	4.00 (0.82)
38	The information (help, on-screen messages, tool-tips, etc.) provided is clear.	3.75 (0.50)
39	It is easy to find the information I need.	4.25 (0.96)
40	The information is effective in helping me complete the tasks and scenarios.	3.50 (0.58)
41	The organization of information on the system screens is clear.	4.50 (0.58)
42	The interface of this system is pleasant.	4.00 (1.41)
43	I like using the interface of this system.	4.25 (0.96)
44	This system has all the functions and capabilities I expect it to have.	4.25 (0.96)
45	Overall, I am satisfied with this system.	4.00 (0.82)
46	I am confident about the results I produced.	4.25 (0.96)

Table 15.5 lists the mean general responses. With the exception of three questions (#36, 38, 40) all the responses agree or strongly agree with the questions, suggesting that the system was generally well organized, capable, easy to use, and well liked by the users. Of course, alternative explanations may be that either the SUT tasks as a whole or the manner of following instructions was overly easy (or both). Only question #36 drew an ambivalent response (neither agree nor disagree) regarding the

utility of the systems error messages. Given that the perceived level of errors generated by the system was low in the first place, this response can be interpreted as either
being inappropriate to the general simplicity of the tasks, or truly insufficient system
responsiveness (feedback). Eye movement evidence provides at least one instance of
support to the latter.

Comparing speed of completion of all users and an expert over all tasks, on average,
the factor of instruction was significant ($F(4, 575) = 5.8375, p < 0.001$). Note that
this comparison was performed over steps performed by all subjects including the
expert on whose tasks the SUTs were based. Unfortunately, the resultant SUTs devi-
ated slightly from the expert's actual actions, and although all four users performed
all of the SUT steps, for the purposes of this analysis, some of the expert's steps must
be considered missing.

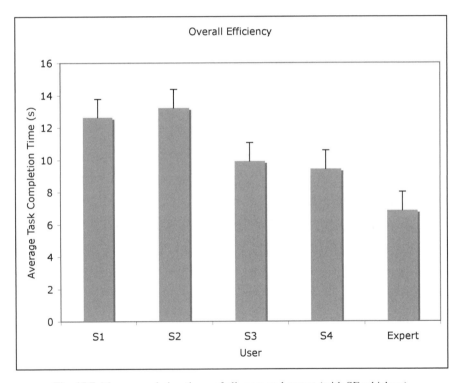

Fig. 15.7. Mean completion times of all users and expert (with SE whiskers).

Mean time to completion for all subjects is shown in Figure 15.7. Posthoc pairwise
t-tests with Bonferroni adjustments indicate significant difference only between each
of S1 and S2 and the Expert ($p < 0.01$).

Performance analysis (time to completion) does not provide very meaningful insights into any potential problematic aspects of the user interface. Rather, it merely indicates that, on average, users given verbal instructions performed the tasks faster than those who read the instructions. Both of these groups of users were slower than the expert who, for analytical consideration of the experiment, had the steps memorized.

To identify particularly troublesome and/or errorprone tasks, all 136 task durations were averaged between users and sorted. The most time-consuming tasks were ones that required typed input; e.g., Step 78 (64.5 s average duration) required entering five typed strings and selection of two drop-down menu items. Steps that did not require as much time generally consisted of single button mouse presses, e.g., Step 23 (Select OK; 2.5 s average duration). Note that the time recorded also included time to read or hear the instruction.

Although most steps did not indicate potential usability effectiveness problems, one action did suggest difficulty: Step 105 (30.5 s average duration, fifth highest). This action requires choosing the Properties dialog from a drop-down menu associated with a graphical box that users were required to overlay atop the DTED map displayed by C2PC. The problem with this instruction is that in order to obtain the correct Properties dialog, the user must right-click on the edge of the graphical element (the box in this case). This is not easy to do if the line width of the element is narrow, as it appears to be by default. This representative user action is shown in Figure 15.8.

Eye Tracking Results. Investigative (ad hoc) analysis of eye movements revealed three usability problems. First, visualization and playback of SUT-002 (Step 68), shown in Figure 15.9, revealed seven errant saccades committed by one of the participants (S_2) during visual search for the Edit button. Figure 15.9 shows the sequence of fixations (#74–#81) just prior to Edit button activation. The six saccades leading up to Edit button activation (between fixations #74–76) are directed toward the menu bar, where, presumably, the participant expected to find the Edit button. In terms of satisfaction, this participant did not report perceiving a problem with the Edit button's unconventional location. However, recall that users may not always report a distracting element during a study (Bojko, 2005). This scanpath interval lasted about 9045 ms (according to the temporal information reported by Tobii's Gaze Plot analysis; this may seem excessive but is probably due to the participant taking time to read instructions that are away from the monitor). Excluding the first and last fixations (618 ms and 1,096 ms, resp.), 7.3 s were lost to this inefficient search (this duration also includes the time to process verbal or written instructions, however).

Second, and also related to performance, feature visibility analysis shows the lack of prominence of an important system status indicator. The interface, shown in Figure 15.10, displays fixational "hotspots" collected from all four participants on this task. The image also shows experimenter-defined AOIs (postfacto), overlaid over specific interface features on a representative image of the interface. Two of these

Fig. 15.8. Example of problematic Properties dialog selection. After two unsuccessful attempts, the user has finally right-clicked on the narrow box outline (at top left of the popup menu).

features, the status icon and the status bar, relate the application's level of activity. When the application is busy, the status icon (a small red stop sign in the toolbar) glows red and the status bar reports the reason for the delay (e.g., "drawing map"). None of the users looked at the status icon, even though they appeared confused by the application's unresponsiveness. Only one user remarked he "could not complete the Overlay task" (ostensibly due to the delay). Figure 15.11 shows descriptive statistics regarding the number of fixations over selected AOIs. Although difficult to read in this compressed version, the second element along the abscissa (status icon) clearly indicates 0 fixations.

Third, area coverage indicators can offer insight into how a user (literally) views the workspace and plans motor actions. In the seventh task, users needed to copy text objects from a source window into a destination window's textbox via two different methods. In once instance, users needed to select the object, then press Add to make the object appear in the destination window. In another instance, users had to click-and-drag the object instead of pressing Add. Area coverage visualizations, ranging

Fig. 15.9. Exemplar errant visual search for Edit button.

from "hotspots" to scanpaths, as shown in Figure 15.12, showed the need for looka-head eye movements to complete the click-and-drag operation. The user had to move the topmost source window out of the way before visually locating the destination location. The Add button clicking action required neither this time-consuming win-dow moving, nor lookahead eye movements. None of the users reported this as a perceived detriment to performance or satisfaction.

15.4.1 Case Study Notes

Traditional usability metrics alone may not fully capture all aspects of a usability study, whether it's due to the design of the experiment, number of participants, pro-cedures, or environment. Control in this case study was quite weak, and thus the efficiency (speed), effectiveness (accuracy), and satisfaction metrics did not reveal a great deal in this instance. Eye tracking metrics, in contrast, identified rather signifi-cant problems with the software suite under inspection.

The key deficiencies in the usability of the system exposed by eye movement metrics all stem from the disregard of rudimentary human factors design principles (consis-tency, visibility, feedback; Norman (1988)) and violation of official user interface

Fig. 15.10. Fixation "hotspots" and selected AOIs.

style guides (e.g., Microsoft's Windows (Microsoft Corporation, 1999)). Specifically, the misplaced Edit button is a violation of visibility because the button's placement is not immediately obvious (nor consistent with Microsoft's traditional menu layout). Affordance was not obvious in the click-and-drag action of various items, especially when other similar actions provided the explicit use of an Add button. Finally, feedback was not provided during the one of the applications' busy states; the simple inclusion of a progress bar would quickly remedy this problem.

In terms of experimental design, this case study is instructive for two reasons. First, it is counterexemplary in that it shows rather poor experimental design. The design suffers from a very small number of participants. Furthermore, the setting is rife with distractions (the conference exhibits' room floor is neither a good example of a controlled lab setting nor an in-field place of work). The detailed nature of the tasks rendered users' actions too simple; very little cognitive effort was probably exerted during execution of the SUTs. There are other numerous problems, all undoubtedly contributing to the inconclusive performance measures obtained. However, the strength of this example lies in its adoption of the KISS principle. Task duration need not be long to expose usability problems. The inconsistent placement of the Edit button,

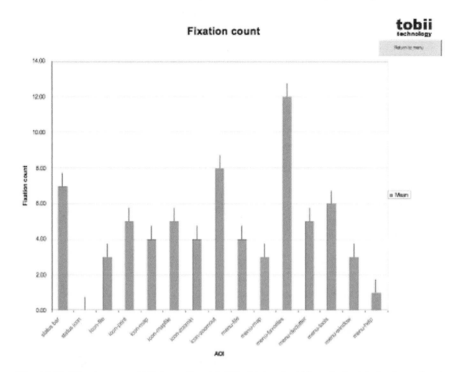

Fig. 15.11. Fixations per AOI (overall, with SE whiskers). Although the labels along the abscissa are barely readable, the noteworthy aspect of this figure is the height of the second element clearly indicating 0 fixations recorded over the status icon.

for example, was identified in only 7.3 s worth of eye tracking data. It's quite possible that had SUT-002 been longer than 2.6 minutes (on average over the four participants), this significant but rather short event may have been missed.

15.5 Desktop Interaction: Gaze-Contingent Fisheye Lens

Motivated in part by classic Fitts' law studies of manual pointing and in part by Zhai et al.'s (1999) gaze cursor warping experiments, Ashmore et al. (2005) evaluated a gaze-contingent fisheye lens as a gaze pointing and selection aid. Applying the Fitts' law model to eye pointing, the study predicted reduced mean selection time through reduction of target distance or expansion of target size, as magnified by the fisheye.

The fisheye lens used in the experiment was the Pliable Display Technology (or PDT) lens, based on the work of Carpendale et al. (1995) and available commercially from Idelix Software.[1] The PDT allows local magnification of data while

---

[1] <http://www.idelix.com>.

Fig. 15.12. Look-ahead saccades for click-and-drag.

keeping a continuous visual connection to the underlying display (Idelix Software Inc., 2004). The lens relies on a geometric transform of data space to display space through a pliable or elastic mesh surface, shown in Figure 15.5(b), enabling content to be stretched from below the surface with lenses of the appropriate height, volume, and shape. Matching data space resolution to magnification factor allows simultaneous target expansion in motor and display space. The resulting visualization provides a "detail-in-context" view of, and interaction with, the display.

Slaving the fisheye to instantaneous gaze coordinates, preliminary anecdotal observations suggested two problems with the gaze-contingent fisheye. First, a continuously slaved fisheye lens appeared to oscillate during gaze selection. Second, being always in front of one's gaze, the fisheye appeared to disrupt preattentive (peripheral) preview benefit; due to its chosen size the lens contracts screen content beneath its shoulder region (see Figure 15.5).

A repeated measures (within-subjects) factorial design was used with the style of fisheye lens interaction as the independent variable, at four levels: nonexistent (no lens; control condition), omnipresent (lens continuously slaved to gaze), MAGIC

(a) Rendered with lens controls.   (b) Underlying mesh representation.

Fig. 15.13. The Pliable Display Technology (PDT) fisheye lens.

(lens appearing only after fixation onset), and Grab and Hold Algorithm, or GHA (as with MAGIC but fixed in place once visible after fixation onset. The MAGIC moniker was borrowed from Zhai et al.'s MAGIC cursor, but unlike the cursor, the appearance of the MAGIC lens was modulated by real-time fixation estimation and was not based on predicted direction of eye movement).

To evaluate the fisheye interaction for gaze pointing, a combined visual search and Fitts' law pointing task was devised with randomly variable target distance (amplitude) and fixed target width when not magnified by the fisheye lens. A 9×9 grid of target boxes served as the targeting stimulus, as shown in Figure 15.14 (left). Animation feedback for gaze targeting and selection was provided by first marking the target box with a large × through its center. Target selection was achieved following a 500 ms fixation (dwell time) within the box boundaries in display space (snap-to-target was not used). Visual confirmation of selection was provided by highlighting the selected box and drawing a red disk centered at the mean fixation position with radius equal to the standard deviation of fixation points collected during the fixation. Following calibration, each participant completed four blocks of 40 trials. Odd-numbered trials required the participant to fixate the central home position box. Even-numbered trials required the participant to search for and select a randomly chosen target from the remaining 80 target boxes (excluding the central target).

Mean selection time (MT; in milliseconds) of participant data pooled per fisheye condition is shown in Figure 15.14 (right) with standard error (SE) whiskers. A standard

Fig. 15.14. Gaze-contingent fisheye lens stimulus and performance results. From Ashmore et al. (2005) © 2005 Canadian Information Processing Society. Reprinted by permission.

weighted-means, one-way ANOVA of correlated samples was performed to examine the main effect of different fisheye conditions. The main effect (fisheye) was significant ($F(3, 1436) = 7.8407, p < 0.01$). Posthoc Tukey pairwise comparisons revealed significant differences between the Omnipresent condition and both GHA and MAGIC conditions ($p < 0.01$) but not the Nonexistent condition. A significant difference was also found between the GHA and Nonexistent conditions ($p < 0.01$) and between the Nonexistent and MAGIC conditions ($p < 0.01$). The GHA and MAGIC conditions gave the fastest performance, with neither significantly better than the other. The Omnipresent and Nonexistent conditions gave the slowest performance and were not significantly different from each other.

Similar analysis was performed for accuracy, as measured by the gaze point's deviation from the target center. Once again the main fisheye effect was significant ($F(3, 1436) = 5.59, p < 0.01$). Posthoc Tukey pairwise comparisons revealed significant differences between the GHA and MAGIC conditions ($p < 0.01$) and between the Nonexistent and MAGIC conditions ($p < 0.01$). The MAGIC and Omnipresent conditions provided the best accuracy, with neither significantly better than the other. The GHA and Nonexistent conditions provided the worst accuracy and were not significantly different from each other.

The most likely reason given for speed improvement outcomes was that both GHA and MAGIC lenses provided the viewer with preview benefit when conducting visual search for the target. Of the three lenses, only the GHA lens was held at a fixed location following fixation onset. If the user's initial accuracy was poor, the target appeared in the distorted lens shoulder, negating any potential target expansion benefits. Although the GHA lens may have offered a speed benefit during visual search, it hindered target selection, thus providing a possible explanation for improved accuracy of only the Omnipresent and MAGIC lenses.

15.5.1 Case Study Notes

This study is a good example of an interactive, or gaze-contingent, eye tracking experiment. It is fairly simple in its construction and control (e.g., simplicity of box targets).

The experiment is also a good example of development of an eye tracking (and PDT fisheye lens) client application. In essence, the resultant application invoked several Software Development Kits (SDKs) to achieve its desired functionality: the Tobii SDK (under Linux), the PDT SDK (from Idelix), as well as less-specialized Application Program Interfaces (APIs) such as OpenGL, Posix threads, and Qt.

It is also worth mentioning that project management in terms of collaborative and revision handling was provided by subversion (svn). Analysis (automated via Postgres MySQL queries) was facilitated by the R analysis language. Eventually, following data collection (the actual experimental trials), the entire analysis was performed by simply typing "make" at the Linux prompt. Paper typesetting was performed by LATEX, the source of which was also held in the same svn repository.

15.6 Summary and Further Reading

This chapter reviewed a number of eye tracking studies performed in Clemson's eye tracking laboratory. Most of the work is described in the associated referenced publications and these ought to be consulted for further details. Indeed, the proceedings that contain these papers or posters hold a wealth of information embodied in similar works. These sources include the proceedings of ETRA (Eye Tracking Research and Applications), SIGCHI, Graphics Interface (GI), EuroGraphics conferences (there are numerous specialized meetings and workshops), and APGV (Applied Perception in Graphics and Visualization). APGV is a fairly recently formed conference, a spinoff from SIGGRAPH, hence is largely concerned with graphics-related work.

Part IV

Eye Tracking Applications

Diversity and Types of Eye Tracking Applications

A wide variety of eye tracking applications exists, which can broadly be described within two categories, termed here as *diagnostic* or *interactive*. In its diagnostic role, the eye tracker provides objective and quantitative evidence of the user's visual and (overt) attentional processes. As an interface modality, the eye tracker serves as a powerful input device that can be utilized by a host of visually mediated applications.

In general, in their diagnostic capacity, eye movements are simply recorded to ascertain the user's attentional patterns over a given stimulus. Diagnostic applications are distinguished by the unobtrusive use of the eye tracking device. In some cases, it may even be desirable to disguise the eye tracker so that potential subjects are not aware of its presence. Furthermore, the stimulus being displayed may not need to change or react to the viewer's gaze. In this scenario, the eye tracker is simply used to record eye movements for posttrial, off-line assessment of the viewer's gaze during the experiment. In this way, eye movement data may be used to objectively corroborate the viewer's point of regard, or overt locus of attention. For example, studies that test the appearance of some aspect of a display, say the location of an advertisement banner on a Web page, may be bolstered by objective evidence of the user's gaze falling on (or missing) the banner under consideration. Typical statistical measurements may include the number of fixations counted over the banner during a five-minute "Web browsing" session. Diagnostic eye tracking techniques are applicable (but not restricted) to the fields of psychology (and psychophysics), marketing/advertising, and human factors and ergonomics.

Equipped with an eye tracker as an input device, an interactive system is expected to respond to, or interact with the user. Interactive applications are therefore expected to respond to the user's gaze in some manner. Such interactive systems may be classified by two application subtypes: selective and gaze-contingent. The latter can be further delineated in terms of display processing, as shown in the hierarchy in Figure 16.1. The archetypical interactive eye tracking application is one where the user's gaze is used as a pointing device. This type of ocular interaction can be considered but one of a set of multimodal input strategies from the system's point

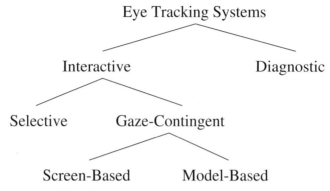

Fig. 16.1. Hierarchy of eye tracking applications.

of view (Hutchinson, 1993; Nielsen, 1993; Schroeder, 1993b, 1993a). An example of a system relying solely on gaze as input has been shown to be an important communication tool for quadriplegics, where the eyes are used for positioning a cursor over an oversized projected keyboard. Using gaze to aid communication has also been explored in multiparty computer-supported collaborative work systems (Vertegaal, 1999). Besides being used as a pointing device, knowledge of the user's gaze may be utilized to alter the display for speed-up purposes, as may be required in the rendering of complex virtual environments (McCormick et al., 1996). Interactive eye tracking techniques are applicable (but not restricted) to the fields of human–computer interaction, visual displays, and computer graphics.

16.1 Summary and Further Reading

A wide variety of (interdisciplinary) eye tracking applications has been and is currently being developed. The examples described and shown here by no means constitute a complete survey of the field, but it is hoped that they provide sufficient motivation to spur further interest in this fascinating area.

Neuroscience and Psychology

A wide assortment of eye tracking studies can be found in the increasingly related fields of neuroscience and psychology. Topics range from basic research in vision science to the investigation of visual exploration in aesthetics (e.g., perception of art). A useful approach to navigating through vast collections of early and contemporary literature is to (for the outset) dissociate high-level cognitive studies from those concerned with a low-level functional view of the brain. In this sense, to use a computational analogy, one can distinguish between the "hardware" (low-level brain circuitry) on which the "software" (high-level cognition) functions. In a complementary view of the apparently disparate disciplines, neuroscience identifies the physiological components that are ultimately responsible for perception. In the context of vision and eye movements, knowledge of the physiological organization of the optic tract as well as of cognitive and behavioral aspects of vision is indispensable in obtaining a complete understanding of human vision.

To illustrate the interdependence of neuroscience and psychology, consider again the scene integration problem (see Chapter 1). Neurophysiological studies clearly identify the visual components involved in dynamic (or active) vision. That is, due to the limited informational capacity of the fovea, the eyes shift from point to point while scanning the visual field. The neuronal organization of retinal cells, which in a sense is the reason for eye movements, is well known. Furthermore, the general organization of foveo–peripheral vision has also been mapped along the magno- and parvocellular pathways leading to deeper regions of the brain and farther still into regions implicated in higher cognitive functions. From psychological observations, we know that humans are aware of a large field of view, even though physiology does not permit a holistic cameralike capture of the entire scene in one exposure. This is the crux of the scene integration problem. Psychologists show us that we are quite adept at maintaining a fairly accurate mental image of the visual scene in front of us. Indeed, various illusory pictures such as the Kanizsa (1976) square show us that we "see" more than what is physically there. The main question of how the brain is able to "piece together" small high-resolution snapshots of the scene remains a mystery.

Although there are many eye-movement related topics examined in the fields of neuroscience and psychology, current neuroscientific trends related to the study of eye movements are briefly discussed touching on exemplary work investigating illusory contours, attention, and brain imaging. Focus is then shifted to psychological investigation of four applied perceptual examples: the study of eye movements in reading, during scene perception (including perception of art and film), visual search, and in natural tasks.

17.1 Neurophysiological Investigation of Illusory Contours

Neuroscientific investigation of vision and in the specific context of eye movements has been covered to a certain extent in previous chapters. Results of this research have led to the identification of various interconnected components of vision, starting from retinal photoreceptors, and (more or less) ending in the cortical regions implicated in low-level vision. Important concepts of retino-geniculate pathways of vision as well as the important lower- and higher-level visual brain regions have been well established. Augmenting traditional neuroscientific investigation with eye tracking devices can lead to further understanding of perceptually puzzling phenomena such as Kanizsa-type illusions.

Logothetis and Leopold (1995) conducted an experiment investigating the neural mechanisms underlying binocular rivalry, the alternating perceptions experienced when two dissimilar patterns are stereoscopically viewed. Single cells were recorded in visual areas V1, V2, and V4 and monkeys reported the perceived orientation of rivaling sinusoidal grating patterns. Monkeys were trained to perform a fixation and an orientation discrimination task. Both tasks required continuous fixation of a small central spot within a 0.8×0.8 degree window. To confirm the location of fixations, eye movements were measured with a scleral coil technique. Logothetis and Leopold suggest that results of this study indicate that awareness of a visual pattern during binocular rivalry arises through interactions between neurons at different levels of visual pathways, and that the site of suppression is unlikely to correspond to a particular visual area, as often hypothesized on the basis of psychophysical observations. Together with earlier psychophysical evidence, the cell types of modulating neurons and their overwhelming preponderance in higher rather than in early visual areas also suggests the possibility of a common neural mechanism underlying binocular rivalry as well as other bistable percepts, such as those experienced with ambiguous figures (e.g., Kanizsa-type illusions and the Necker cube). That is, perception of illusory contours may be physiologically related to stereoscopic perception.

17.2 Attentional Neuroscience

An important problem peculiar to eye tracking studies is that of the dissociation of visual attention from the point of regard. That is, most of the time, we can assume that one's visual attention is associated with the point of fixation. This is usually referred to as overt attention, because it is the component of visual attention (when associated with foveal vision) that can be measured overtly (i.e., by an eye tracker). However, it is certainly possible to fix one's gaze at a specific point, and yet move one's attention to a nearby region. Astronomers do this fairly regularly when looking for faint stars or star clusters with the naked eye. The Little Dipper is a good example of star clusters that is found more easily when one looks for it "off the fovea".

The dissociation of attention from ocular fixation poses a problem for eye tracking researchers. When examining a scanpath over a visual stimulus, we can often say that specific regions were "looked at", perhaps even fixated (following analysis of eye movements), however, we cannot be fully confident that these specific regions were fully perceived. There is (currently) simply no way of telling what the brain was doing during a particular visual scan of the scene. Ideally, we would have to not only record the point of one's gaze, but also of one's brain activity. Research that combines eye tracking with traditional neuroscientific paradigms offers the dual benefit of monitoring brain activity as well as oculomotor function.

Investigating neuronal activity related to fixational eye movements, Snodderly et al. (2001) used a double Purkinje image eye tracker (2–3 minarc resolution; 100 Hz sampling rate) to record the position of a macaque monkey's eye when fixating a light-emitting diode for 5 seconds. Action potentials were recorded from neurons in area V1 of three adult female macaque monkeys that were trained to hold visual fixation. Snodderly et al. show that responses of V1 neurons to fixational eye movements are specific and diverse. Some cells are activated only by saccades, others discharge during drift periods, and most show a mixture of these two influences. Three types of eye movement activation were found: (1) "saccade-activated cells" discharged when a fixational saccade moved the activating region onto the stimulus, off the stimulus, and across the stimulus; (2) "position/drift cells" discharged during the intersaccadic (drift) intervals and were not activated by saccades that swept the activating region across the stimulus without remaining on it; and (3) "mixed cells" fired bursts of activity immediately following saccades and continued to fire at a lower rate during intersaccadic intervals. The patterns of activity reflect the interactions among the stimulus, the receptive-field activating region, the temporal response characteristics of the neuron, and the retinal positions and image motions imparted by eye movements. The diversity of the activity patterns suggests that during natural viewing of a stationary scene some cortical neurons are carrying information about saccade occurrences and directions whereas other neurons are better suited to coding details of the retinal image.

Snodderly et al. report that the two components of fixational eye movements, saccades and drifts, activate different subpopulations of V1 neurons in distinctive ways. Saccade-activated and mixed cells fire bursts of spikes when a fixational saccade moves the activating region onto the stimulus, off the stimulus, or across the stimulus. The sign of contrast (light or dark) is unimportant. These characteristics imply that the burst responses are conveying rather crude information about the details of the stimulus. For example, if an appropriately oriented stimulus contour activated the neuron following a saccade, it would be difficult to determine whether the contour remained in the activating region (saccade moved region onto contour) or was outside it (saccade moved region across contour). This ambiguity would have particular relevance when viewing complex natural images, because the neuronal discharge could be evoked either by a stimulus feature crossed by the activating region during the saccade, or by a completely different feature on which the region landed at the end of the saccade. Bursts of spikes fired by saccade and mixed cells following fixational saccades may suggest that the bursts are important sources of information about the visual scene. An alternative role for the saccade neurons might be to participate in the suppression of visual input associated with saccades. Snodderly et al. argue that, theoretically, saccade neurons could participate in saccadic suppression by inhibiting other neurons that carry stimulus information, or by adding noise to the signal, thereby raising thresholds and making stimulus events undetectable at the times of saccades. This role would make the saccade neurons the source of saccadic suppression.

According to Snodderly et al., the position/drift neurons must play a quite different set of roles that are complementary to those of the other eye movement classes. Because the position/drift neurons do not respond to saccades, they may be spared the potentially detrimental effects of saccade-related activity. They signal accurately the position of stimulus features on the retina, and in many cases the sign of contrast. Thus, their activity could in principle be the basis for a reconstruction of the image.

Involved in attentional shifting behavior, the prefrontal (PF) cortex has long been thought to be central to the ability of choosing actions appropriate not only to the sensory information at hand but also according to the situation in which it is encountered (Asaad et al., 2000). Recent studies, reviewed by Asaad et al., indicate that the specific sensory, motor, and cognitive demands of the task (the behavioral context) can be an important factor in determining PF neural responses. For example, neural activity to an identical visual stimulus can vary as a function of which portion of that stimulus must be attended or with the particular motor response associated with it. Damage to the PF cortex of humans and monkeys tends to produce impairments when available sensory information does not clearly dictate what response is required. For example, PF lesions impair spatial delayed response tasks in which a cue is briefly flashed at one of two or more possible locations and the subject must direct an eye movement to its remembered location.

Asaad et al. performed an eye tracked neurophysiological experiment to explore the role of the PF cortex. Subjects were two rhesus monkeys (with immobilized heads). One animal was implanted with an eye-coil to monitor eye movements, and an infrared monitoring system from ISCAN was used for the second animal. Eye position was monitored at 100 Hz in both animals. Microsaccades and saccades were detected using a simple velocity threshold set at four times the standard deviation of the signal derived from the fixation period. Neural recording sites were localized using magnetic resonance imaging (MRI). Recording chambers were positioned stereotaxically over the left or right lateral prefrontal cortices of each animal, such that the principal sulcus and surrounding cortex, especially the ventrolateral PF cortex, was readily accessible. Recordings were made using arrays of eight durapuncturing, tungsten microelectrodes. Activity of up to 18 individual neurons could be recorded simultaneously in any given session. Signal from 210 neurons was recorded from the left lateral PF cortex of one monkey and 95 neurons from the right lateral PF cortex of the other. Monkeys performed an object memory task (delayed match-to-sample), an associative task (conditional visuomotor task), and a spatial memory task (spatial delayed response). The first two tasks shared common cue stimuli but differed in how these cues were used to guide behavior, whereas the latter two used different cues to instruct the same behavior. All three required the same motor responses. The associative task required the animals to associate a foveally presented cue stimulus with a saccade either to the right or left. The object task used the same cue stimulus as the associative task; however, in this case, they needed only to remember the identity of the cue and then saccade to the test object that matched it. Conversely the spatial task used small spots of light to explicitly cue a saccade to the right or left and required the monkeys to remember simply the response direction.

Asaad et al. report that most of the 305 recorded neurons displayed a task-dependent change in overall activity, particularly in the fixation interval preceding cue presentation. Results show that for many PF neurons, activity was influenced by the task being performed. This influence included changes in their baseline firing rates, modulations of neuronal activity related to particular stimuli and behavioral responses, and difference in their firing rate profiles. Asaad et al. suggest that results indicate the formal demands of behavior are represented within PF activity and thus support the hypothesis that one PF function is the acquisition and implementation of task context and the "rules" used to guide behavior.

17.3 Eye Movements and Brain Imaging

Recent investigations of eye movements and functional brain imaging simultaneously examine readings from an eye tracker and from a device that images brain activity. For example, Gamlin and Twieg (1997) have embarked on designing a combined visual display and eye tracking system for high-field fMRI (functional Magnetic Resonance Imaging) studies. The proposed project is a collaboration between a neuroscientist and biomedical engineer. The technological goal is the design, development, and implementation of a combined high-resolution binocular display and eye tracking system for use in fMRI studies of the human brain. The scientific goal of the project is the study of neural control of saccadic eye movements, allowing investigation of stereopsis and depth perception, as well as permitting oculomotor studies of smooth pursuit, optokinetic nystagmus, vergence, and accommodative eye movements.

Systems that marry functional brain imaging with eye tracking can be used to at least corroborate a subject's fixation point while simultaneously recording cortical enhancement during attentional tasks. Presently, possibly due to prohibitive cost, combined eye tracking and brain imaging equipment is not widespread, although such devices are beginning to appear; e.g., see Figure 17.1.

Fig. 17.1. Example of eye tracking fMRI scanner. Courtesy of SensoMotoric Instruments (SMI), Needham, MA <http://www.smiusa.com>. Reproduced with permission.

Employing a combined brain imaging and eye tracking device, possibly similar to the one shown in Figure 17.1, Özyurt et al. (2001) used an fMRI device to compare the neural correlates of visually guided saccades in the step and gap paradigms. During task performance, saccadic eye movements were recorded with an MR-Eyetracker. Subjects viewed stimuli that were projected onto a screen attached to the front of the MR scanner. A block-design fMRI was used with alternating blocks of step, rest, and gap trials. Results from the study by Özyurt et al. indicate significant task-related activity in striate and extrastriate cortex, the frontal eye fields, the supplementary motor area, parietal cortex, and angular gyrus, the frontal operculum, and the right prefrontal area 10. This type of research helps identify functional brain structures implicated in attentional behavior.

17.4 Reading

Perhaps the first well-known applied uses of eye trackers in the study of human (overt) visual attention were those conducted during reading experiments. An excellent book on eye movements was collected by Rayner (1992), an influential researcher in eye movements and reading. Rayner's collection of articles contains exemplary research on reading and on scene perception. Rayner's (1998) article gives a rather comprehensive survey of eye tracking applications, reviewing studies of eye movements in reading and other information-processing tasks such as music reading, typing, visual search, and scene perception. The major emphasis of the review is on reading as a specific example of cognitive processing.

Rayner (1998) reviews a good deal of previous work on eye movements synthesizing a large amount of information gleaned from over 100 years of research. Although the reader is referred to Rayner's article for the complete review, three interesting examples of eye movement characteristics during reading are given here. First, eye movements differ somewhat when reading silently from reading aloud: mean fixation durations are longer when reading aloud or while listening to a voice reading the same text than in silent reading. Second, when reading English, eye fixations last about 200–250 ms and the mean saccade size is seven to nine letter spaces (see below). Third, eye movements are influenced by textual and typographical variables; e.g., as text becomes conceptually more difficult, fixation duration increases and saccade length decreases. Factors such as the quality of print, line length, and letter spacing influence eye movements. There is of course a good deal more that has been learned, however, here the methodology behind such discoveries is what is of primary interest. Below, three main experimental paradigms used in eye tracking experiments are discussed.

Three experimental paradigms, the *moving window*, *boundary*, and *foveal mask*, have been developed to explore eye movements. Although first developed for reading studies, these paradigms have since been adapted to other contexts such as scene perception (see below). In the moving window paradigm, or *gaze-contingent display change* paradigm, developed by McConkie and Rayner (1975), a window is sized to include a number of characters (e.g., 14) to the left and right of a fixated word. For example, the sentence

```
the quick brown fox jumped over the lazy dog
```

is presented as follows in four subsequent temporal instances (note the change of word   fox   to   cat;   the   asterisks   indicate   fixation   locations).

```
xhe quick brown xxxxxxxxxxxxxxxxxxxxxxxxxxxx
               *
xxxxxxxxk brown fox jumxxxxxxxxxxxxxxxxxxxxxx
                    *
xxxxxxxxxxxxxxxxxxcat jumped ovexxxxxxxxxxxxxx
                         *
xxxxxxxxxxxxxxxxxxxxxxxxxxxxd over the laxxxxxx
                                        *
```

The assumption with this technique is that when the window is as large as the region from which the reader can obtain information, there is no difference between reading in that situation and when there is no window. A related but reverse method, developed by Rayner and Bertera (1979) (see also Bertera and Rayner (2000)), places a foveal mask over a number of fixated characters (e.g., seven):

```
the qxxxxxxxown fox jumped over the lazy dog
    *
the quick brown cat jumpedxxxxxxxhe lazy dog
                          *
```

This situation creates an artificial foveal scotoma and eye movement behavior if the situation is quite similar to the eye movement behavior of patients with real scotomas (Rayner, 1998).

In the boundary technique, developed by Rayner (1975), the stimulus changes as fixation crosses a predefined boundary. Rayner used eye movements to investigate reading because he found tachistoscopic methods inadequate: tachistoscopic (strobelike) presentation of letters and words relies on the presentation of material for very brief exposures to exclude the possibility of an eye movement during the presentation. Prior to eye tracked reading studies, this method was often thought of as being analogous to a single fixation during reading. Based on his and others' research, Rayner argued that what subjects report from a tachistoscopic presentation cannot be taken as a complete specification of what they saw. The argument for eye movement recording over tachistoscopic displays carries over to scene perception and is discussed further in the next section.

In an example of the boundary technique below, a single critical target word is initially replaced by another word or by a nonword. The boundary paradigm allows the experimenter to be more diagnostic about what kind of information is acquired at different distances from fixation. In this technique, a word or nonword letter string is initially presented in a target location. However, when the reader's eye movement crosses an invisible boundary location, the initially presented stimulus is replaced by the target word. The amount of time that the subject looks at the target word is influenced by the relationship between the initially presented stimulus and the target word and the distance from the launch site to the target location. The fixation time on the target word thus allows the experimenter to make inferences about the type of information acquired from the target location when it was in parafoveal vision. For example, the word fox is changed to cat when the eyes cross the boundary.

```
the quick brown fox jumped over the lazy dog
             *       |
the quick brown cat jumped over the lazy dog
                 |                   *
```

The assumption is that if a reader obtains information from the initially presented stimulus, any inconsistency between what is available on the fixation after crossing the boundary and with what was processed on the prior fixation is registered in the fixation time on the target word.

Experiments using the above gaze-contingent techniques have shown, generally, (1) that readers typically acquire the visual information necessary for reading during the first 50–70 ms of a fixation, and (2) that a serial scan of letters in foveal vision does not occur. Furthermore, studies using these techniques consistently indicate that the size of the perceptual span is relatively small (e.g., for readers of alphabetical orthographies, such as English, Dutch, or French, the span extends from 3–4 letters to the left of fixation to about 14–15 letter spaces to the right of fixation), with a still smaller word identification span (Rayner, 1998).

Experiments in reading can lead to descriptions of individuals' reading strategies, or perhaps even to suggestions for improvement of one's reading strategy. Previous findings have shown that when an orienting visual response (eye movement) is made to a target pair consisting of two neighboring but separated elements, the first saccade lands in between the two elements. This was called the *global effect* by Findlay (1992).[1] In examining the global effect in reading, O'Regan (1992) suggests that an "optimal" viewing position may exist, where the time it takes to recognize the word is minimal. Due to oculomotor constraints or scanning strategies, the eye does not always land at the optimal spot. O'Regan discusses a combination of two observed strategies and tactics for reading, as shown in Table 17.1. A double fixation involves a second fixation that is inversely proportional (in duration) to the first one. If far from optimal position, the fixation is very short (100–150 ms). Single fixations may be long (up to 300 ms). The optimal viewing position phenomenon is weak in reading, but strong in single word experiments. That is, it is not known what the "special phenomenon" is in reading that accounts for the weakening of the optimal viewing position phenomenon. It may depend on the reader's strategy: risky or careful. It is not known whether readers can choose to read faster or slower.

The various types of gaze-contingent word display change techniques have had a major impact on reading research and scene perception because they allow the experimenter to control the information available to the subject at any point in time. Such experimental control over the nature of timing of the available information has

---

[1] The effect may also be referred to as the *center of gravity* effect.

Table 17.1. Reading strategies/tactics.

Tactic	"Careful" strategy	"Risky" strategy
Single fixation	One fixation in the critical region	One fixation in the critical region
Double fixation	One fixation before critical region, one after (or the reverse–one fixation after and one backwards fixation before critical region)	Two fixations in the critical region

enabled researchers to draw interesting conclusions about on-line aspects of reading and scene perception.

17.5 Scene Perception

Although certain reading patterns are easily recognized (e.g., left-to-right, top-to-bottom for English readers, or right-to-left for Hebrew), no apparent strategies for scene viewing have been easily discerned. Contrary to reading, there appears to be no canonical scanpath for particular objects (i.e., there is no particular "right way" to look at objects; Kennedy (1992)). Kennedy suggests that the reading task is composed almost exclusively of saccades, whereas picture viewing is composed of shifts, pursuits, and drifts. There may be context differences at play. Continuing the debate about context effects for scenes and sentences, Kroll (1992) states that although there may be similarities between the two tasks, the tasks are very different. Eye movements in reading are to a large extent driven by the well-known, practiced task, however, we don't know what the "glue" is that holds the scene together. Kroll states,

> One of the common problems in this research is to develop a set of tasks that will allow us to uniquely locate the interaction of context with object recognition over time.

Rayner (1998) recounts the traditionally held belief that examining the fine details of eye movements during scene perception is a high-cost, low-yield endeavor. Experiments using tachistoscopic presentations and eye movement recordings have led to the conclusion that participants get the gist of a scene very early in the process of looking, sometimes even from a single brief exposure. Thus it has been advocated that the gist of the scene is abstracted on the first few fixations, and the remainder of the fixations on the scene are used to fill in details. Such findings give rise to the question of the value gained from information obtained from detailed eye movement analyses as people look at scenes. Rayner reviews several findings that support the contention that important conclusions about temporal aspects of scene perception can be obtained from eye movement analysis. He recounts the argument put forth by Loftus (1981) and Rayner and Pollatsek (1992) stating that tachistoscopic studies have not shown conclusively that they reveal a perceptual effect rather than the outcome of memory processes or guessing strategies.

Loftus (1981) presented results of a masked tachistoscopic study which suggest a model of picture encoding that incorporates the following propositions: (a) a normal fixation on a picture is designed to encode some feature of the picture; (b) the duration of a fixation is determined by the amount of time required to carry out the intended feature encoding; and (c) the more features are encoded from a picture, the better the recognition memory will be from the picture. A major finding of Loftus' experiments was that with exposure time held constant, recognition performance increased with increasing numbers of fixations. When eye fixations are simulated tachistoscopically and their durations experimentally controlled, all traces of this phenomenon disappear. Moreover, Loftus' experiments suggest that within a fixation, visual information processing ceases fairly early; that is, acquired information reaches asymptote soon after the start of a fixation.

An eye fixation has the very salient property that it shifts the gaze to a new place in the picture. The problem identified with tachistoscopic exposures, which were meant to simulate fixations to new places, is that there is no guarantee that this occurred; it is entirely possible that subjects were simply holding their eyes steady throughout all tachistoscopic flashes. From an eye tracking experiment, Loftus draws the argument that given more places to look at in the picture, more information can be acquired from the picture. Additional (tachistoscopic) flashes are only useful insofar as they permit acquisition of information from additional portions of the picture. Information pertinent to subsequent recognition memory seems to be acquired only from the small $2° \times 3°$ foveal region during a given fixation, and a fixation is useful only to the degree that it falls on a novel place in the picture.

The fixational perceptual span in scene perception mirrors that for reading with one important difference: meaningful information can be extracted much farther from fixation in scenes than in text (Rayner, 1998). Objects located within about 2.6° from fixation are generally recognized, but recognition depends to some extent on the characteristics of the object. Qualitatively different information is acquired from the region 1.5° around fixation than from any region farther from fixation. At high eccentricities, severely degraded information yields normal performance. This suggests that low-resolution information is processed in the more peripheral parts of the visual field, whereas high-resolution information is processed in foveal vision. High spatial frequency information is more useful in parafoveal and peripheral vision than low spatial frequency information.

Rayner and Pollatsek (1992) concede that much of the global information about the scene background or setting is extracted on the initial fixation. Some information about objects or details throughout the scene can be extracted far from fixation. However, if identification of an object is important, it is usually fixated. The work discussed in Rayner and Pollatsek's paper indicates that this foveal identification is aided significantly by the information extracted extrafoveally. Rayner and Pollatsek conclude that it is necessary to study eye movements to achieve a full understanding of scene perception. They argue that if the question of interest is how people process scenes in the real world, understanding the pattern of eye movements will be an important part of the answer.

Rayner (1998) summarizes a number of other findings and claims, some controversial, eventually asserting that given the existing data, there is fairly good evidence that information is abstracted throughout the time course of viewing a scene. Rayner concludes that whereas the gist of the scene is obtained early in viewing, useful information from the scene is obtained after the initial fixations.

According to Henderson and Hollingworth (1998), there are at least three important reasons to understand eye movements in scene viewing. First, eye movements are critical for the efficient and timely acquisition of visual information during complex visual-cognitive tasks, and the manner in which eye movements are controlled to service information acquisition is a critical question. Second, how we acquire, represent, and store information about the visual environment is a critical question in the study of perception and cognition. The study of eye movement patterns during scene viewing contributes to an understanding of how information in the visual environment is dynamically acquired and represented. Third, eye movement data provide an unobtrusive, on-line measure of visual and cognitive information processing. Henderson and Hollingworth list two important issues for understanding eye movement control during scene viewing: *where* the fixation position tends to be centered during scene viewing, and *how long* the fixation position tends to remain centered at a particular location in a scene.

Henderson and Hollingworth (1998) review past results indicating that the positions of fixations within a scene are nonrandom, with fixations clustering on informative scene regions. However, the specific effect of semantic informativeness beyond that of visual informativeness on fixation position is less clear. Several metrics can be used to evaluate the relative informativeness of scene regions: at a macro-level analysis, *total time* that a region is fixated in the course of scene viewing (the sum of the durations of all fixations in that region); this measure is correlated with the number of fixations in that region. At a micro-level analysis, several commonly used measures include *first fixation duration* (the duration of the initial fixation in a region), *first pass gaze duration* (the sum of all fixations from first entry to first exit in a region), and *second pass gaze duration* (the sum of all fixations from second entry to second exit in a region). Generally, first pass gaze durations are longer for semantically informative (i.e., inconsistent) objects. Semantically informative objects also tend to draw longer second pass and total fixation durations. The influence of semantic informativeness on the duration of the very first fixation on an object is less clear. That is, scene context has an effect on eye movements: fixation time on an object that belongs in a scene is less than fixation time on an object that does not belong (Rayner, 1998). However, it is not clear whether the longer fixations on objects in violation of the scene reflect longer times to identify those objects or longer times to integrate them into a global representation of the scene (it could also reflect amusement of the absurdity of the violating object in the given context).

Henderson (1992) asks what the influence is of the contextual constraint provided by a predictive scene on the identification of its constituent objects. For example, does the context serve to facilitate identification procedures for objects consistent with that scene? Concretely, can a cow be identified more accurately and/or more quickly if it is viewed in a farm scene rather than in a kitchen scene? Henderson summarizes the predominant view of the relation between object and scene identification as the *schema hypothesis*. Although there are several variations on the schema theme, several commonalities define the hypothesis. According to the schema hypothesis, a memory representation of a prototypical scene is quickly activated during scene viewing, and is used to develop expectations about likely objects. These expectations then influence the object identification processes. For example, under the schema hypothesis, the identification of a cow in a farm scene involves (1) quickly recognizing that the scene is an exemplar of the category "farm scene", (2) accessing from memory the schema for a farm scene, (3) using the information stored with the schema to generate "cow" and other object candidates likely to be found in a farm scene (and possibly their canonical spatial relationships), and (4) using the knowledge that a cow is likely in such a scene to aid object identification processes when the cow is encountered. The above description suggests a serial model, however, this need not be a central assumption of the schema hypothesis. Henderson cautions against the schema hypothesis and suggests the use of eye tracking to search for a better model of scene perception.

Henderson (1992) offers two criticisms of the eye movement paradigm. First, it is likely that global measures of fixation time, such as the total time spent on an object during the course of scene viewing, and the gaze duration on an object (the time of all initial fixations on an object prior to leaving that object for the first time) reflect postidentification processes. Thus, it is likely that gaze duration in scene processing reflects other processes beyond object identification. Henderson suggests that the preferred fixation measure is the true first fixation duration, or the duration of time from the initial landing of the eyes on an object until the eyes move to any other location, including another location on the object. Second, the basic premise of the eye movement paradigm is that the results will reflect normally occurring visual-cognitive processes because subjects can view scenes in a natural manner. However, unlike reading, where the overall task is arguably transparent, subjects must be given an orienting task when they view a scene. Unfortunately, viewing behavior and eye movement patterns change as a function of the viewing task given to the subject (Yarbus, 1967). One way to address the orienting task issue would be to give subjects a task that did not force the creating of a coherent memory representation for the scene, and look for similar scene context effects on fixation time across tasks.

17.5.1 Perception of Art

A particularly interesting subset of scene perception studies is the examination of gaze over a specific set of contextual images, namely art. As observed by Yarbus (1967), a viewer's intent influences eye movements and fixations over a scene. Eye movements over art have further refined the scope of these studies, examining differences in how trained viewers search for meaning and aesthetic qualities in fine art pieces.

The first systematic exploration of fixation positions in scenes was reported by Buswell (1935) (as cited by Henderson and Hollingworth (1998)), who asked 200 participants to look at 55 pictures of different types of artwork under a variety of viewing instructions. An important result was that fixation positions were found to be highly regular and related to the information in the pictures (Henderson & Hollingworth, 1998). For example, viewers tended to concentrate their fixations on people rather than on background regions when examining *Sunday on the Island of La Grande-Jatte* by Georges Seurat. These data thus provided some of the earliest evidence that eye movement patterns during complex scene perception are related to the information in the scene, and by extension, to perceptual and cognitive processing of the scene. Buswell concluded that:

> Eye movements are unconscious adjustments to the demands of attention during a visual experience. The underlying assumption in this study is that in a visual experience the center of fixation of the eyes is the center of attention at a given time.

Another example of eye movement over art studies, by Molnar (1981) (as reported by Solso (1999)), shows small differences in scanpaths between groups of subjects viewing artwork for its semantic meaning or for its aesthetic appeal. Remarkably, however, both sets of scanpaths are very similar in terms of fixated image features. In Rembrandt's *Anatomy Lesson*, fixations made by two groups of fine art students, given different sets of questions pertaining to the painting, appear to coincide on important elements even though the order of fixations may differ. In this case, the important elements are those of the faces of the anatomy professor and his students in the painting.

Further analyses of eye movements collected over art pieces reported by Molnar (as cited in Solso (1999)) seem to suggest a general rule: more complex pictures produce shorter eye fixations than less complex forms. For example, in comparison of classical works of the high Renaissance to those of the mannerist and baroque periods, classical art produces eye movements that are large and slow, reflecting the expansive nature of that style, whereas baroque paintings involve small and quick eye movements, reflecting the dense animated character of that form. Baroque paintings were judged by art experts as more complex than classical paintings. It may be that complex art, which is densely packed with details, demands that attention be given to a large number of visual elements. This demand can be satisfied by allocating shorter fixation times to each feature. Simpler works may contain far fewer features vying for attention, and so may allow more time allocation per feature, resulting in longer fixations per feature.

A recent large-scale eye tracking study of art was conducted by Wooding (2002). An automated eye tracker was left running in a room of the National Gallery, London, as part of the millennium exhibit: "Telling Time." In three months, eye movements were successfully collected from 5638 subjects while they viewed digitized images of paintings from the National Gallery collection. Because a composite representation of eye movements from so many subjects posed a problem, Wooding devised a *fixation map* method of analysis, which might be descriptively termed a landscape or terrain map of fixations, and is in fact similar to the *landscape map* developed independently by Velichkovsky et al. (1996). The value at any point on the map indicates the height or amount of a particular property at that point (e.g., the number of fixations). An example of a fixation map and subsequent visualization of composite fixated regions are shown in Figures 17.2 and 17.3.

Future eye tracking studies over art, be they comparative or of a mass scale, will undoubtedly lead to further insights of human perception of this particularly pleasing set of images. In a similar goal, but being approached from a different starting point, eye movement and general perceptual principles are currently being applied to the generation or creation of art by computers. A recent collaborative gathering of computer graphics and other scientists, such as representatives from perceptual and cognitive fields, met in Utah (McNamara & O'Sullivan, 2001). There, perceptual principles and eye tracking methodologies were discussed as a possible aid to the creation of more aesthetically pleasing or more interactive works of computer graphics and art.

17.5.2 Perception of Film

Another interesting form of artistic media is film. In a sense, film is a dynamic form of art. An interesting example of an eye tracking film study is given by d'Ydewalle et al. (1998). d'Ydewalle et al. distinguish three levels of film editing errors in sequencing successive shots. First-order editing errors refer either to small displacements of the camera position or to small changes of the image size, disturbing the perception of apparent movement and leading to the impression of jumping. Second-order editing errors follow from a reversal of the camera position, leading to a change of the left–right position of the main actors (or objects) and a complete change of the background. With third-order editing errors, the linear sequence of actions in the narrative story is not obeyed. d'Ydewalle et al.'s experiment shows that there is an

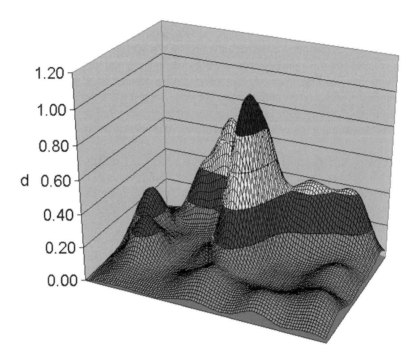

Fig. 17.2. Fixation map from 131 subjects viewing Paolo Veronese painting *Christ Addressing a Kneeling Woman*. From Wooding (2002) © 2002 ACM, Inc. Reprinted by permission.

increase of eye movements from 200 to 400 ms following both second- and third-order editing errors. Such an increase is not obtained after a first-order editing error, suggesting that the increase of eye movements after second- and third-order editing errors is due to postperceptual cognitive effects.

17.6 Visual Search

The question of how humans perceive the visual scene through the movements of the eyes, in the context of more natural or free tasks such as picture viewing (which is significantly different from the task of reading), can generally be modeled by the process known as visual search. Visual search, in general, refers to the process of visually scanning a scene and forming a conceptual "image" or notion of the scene as assembled by the brain. In comparison to reading, there have not been nearly as many studies dealing with visual search (Rayner, 1998).

When eye movements are recorded during extended search, fixations tend to be longer than in reading. However, there is considerable variability in fixation time and saccade length as a function of the particular search task (Rayner, 1998). Specifically,

(a) Original image.

(b) Visualization of important regions.

Fig. 17.3. Sample fixations from 131 subjects viewing Paolo Veronese *Christ Addressing a Kneeling Woman* © National Gallery, London, with annotations © IBS, University of Derby, UK. From Wooding (2002) © 2002 ACM, Inc. Reprinted by permission.

visual search tasks vary widely, and tasks in which eye movements have been monitored consist of at least the following: search (a) through text or textlike material, (b) with pictorial stimuli, (c) with complex arrays such as X-rays, and (d) with randomly arranged arrays of alphanumeric characters or objects. Because the nature of the search task influences eye movement behavior, any statement about visual search and eye movements needs to be qualified by the characteristics of the search task.

Henderson and Hollingworth (1998) list factors that may vary from study to study: image size (usually measured in visual angle), viewing task (e.g., later recognition memory task, image preference, counting nonobjects, visual search, "free viewing" (it is well known that viewers place their fixations in a scene differently depending on viewing task), viewing time per scene (ranging from very short, tachistoscopic to longer durations, e.g., on the order of 50 ms–10 s or even 30 minutes (Yarbus, 1967)), image content (e.g., artwork, "natural scenes", human faces), and image type (e.g., highly artificial and regular such as sine wave gratings, color, monochrome, grayscale, or full-color computer graphics imagery). These factors could each produce main effects and could also interact with each other in complex ways to influence dependent measures of eye movement behavior such as saccadic amplitudes, fixation positions, and fixation durations.

As demonstrated by Yarbus (1967) and then by Noton and Stark (1971a, 1971b), eye tracked scanpaths strongly suggest the existence of a serial component to the picture viewing process. However, serial scanpaths do not adequately explain the brain's uncanny ability to integrate holistic representations of the visual scene from piecemeal (foveal) observations. That is, certain perceptual phenomena are left unaccounted for by scanpaths, including perception of illusory images such as the Kanizsa (1976) figures or the Necker (1832) cube. Although scanpaths cast doubt on a purely Gestalt view of visual perception, it would seem that some sort of holistic mechanism is at work which is not revealed by eye movements alone. Models of visual search attempt to answer this dilemma by proposing a parallel component, which works in concert with the serial counterpart exhibited by eye movements.

In visual search work, the consensus view is that a parallel preattentive stage acknowledges the presence of four basic features: color, size, orientation, and presence and/or direction of motion (Doll, 1993; Wolfe, 1993). Todd and Kramer (1993) suggest that attention (presumably in the periphery) is captured by sudden onset stimuli, uniquely colored stimuli (to a lesser degree than sudden onset), and bright and unique stimuli. There is doubt in the literature whether human visual search can be described as an integration of independently processed features (Van Orden & DiVita, 1993).

Visual search, even over natural stimuli, is at least partially deterministic, rather than completely random (Doll et al., 1993). Determinism stems from either or both of two kinds: the observer's strategy determines the search pattern (as in reading), and/or the direction of the next saccade is based on information gained through peripheral vision about surrounding stimuli. Features that are likely to be fixated include edges, corners, spatially high-frequency components, but not plain surfaces.

A theory that consolidated the serial and parallel counterparts of visual attention was put forth by Treisman (1986) (see also Treisman and Gelade (1980)). Treisman and her colleagues reported on the "pop-out" effect; that is, a visual target pops out from a field of similar but distinct distractors, seemingly drawing attention to itself (see Figure 17.4). This observation led to the proposition that the visual system can search for certain stimulus properties in parallel over the whole visual field, a preattentive process. Other target items that do not pop out must be searched for by the observer in some (usually serial) order-focused attentional processing. The difference between preattentive and postattentive modes of processing can be demonstrated experimentally via paradigms designed to elicit visual search. In a typical visual search task, the observer is shown an array of items and asked to search for a specific target among the field of distractors. Reaction time is usually measured when reporting the presence or absence of a target. If reaction time does not vary with the number of items, the whole array is thought to be searched in parallel, or preattentively. If reaction time increases with the number of items, the array is most likely being scanned item by item in a serial process that is depicted by scanpaths. Treisman proposed the Feature Integration Theory (FIT) unifying these processes.

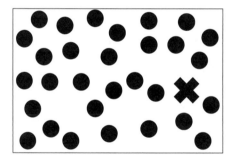

Fig. 17.4. Example of "pop-out" effect.

The main tenet of FIT is that certain features pop out because they are explicitly represented by low-level vision. This is consistent with known retinotopic maps of cortical cells tuned to basic stimulus properties such as color, orientation, and size. Conjunctions of features, however, require serial search because they do not form part of this initial representation. Thus, according to this model, visual processing can be described by three broad stages: (1) feature extraction, (2) feature binding, and (3) object representation.

Feature integration theory is a theory of how elementary visual features are attentionally bound together to construct unitary perceptual objects. The first (parallel) stage of processing involves multiple, retinotopic master maps of locations of elementary features, showing where all feature boundaries are located, but not what those features are. Maps correspond to feature attributes such as color, orientation, etc. According to FIT, perceptual objects are constructed by attention to a specific location, binding together the simultaneous activity at the location in all feature maps. Thus, attention can be thought of as a "glue" that integrates separated feature attributes at a particular location so that the conjunction is perceived as a unified whole.

FIT is a useful theory because it adequately explains both serial and parallel components of visual search. Parallel search is supported by FIT because targets defined by elementary properties are available in feature maps. Serial search is supported ince search for conjunctions requires focused attention to bind together features in separate maps. However, although FIT can explain search performance fairly well, particularly over simple stimuli (e.g., search for Q in a field of Os), it is not clear whether FIT generalizes to more complex stimuli such as natural imagery (e.g., aerial photographs).

Feature integration theory is currently held to be a useful heuristic starting point for theoretical treatments of visual search although it is widely recognized to be over-simplified (Findlay & Gilchrist, 1998). Searches for more sophisticated accounts have taken various forms. Wolfe (1994) has pointed out the weakness of the assumption that search must be either serial or parallel and developed a model of the way in which interactions between serial and parallel processes could occur. A particular problem of FIT is that it suggests that although preattentive feature processes can perform feature searches in parallel, attention from item to item is required for all other (serial) searches. Conjunction searches, however, appear to be too efficient to be explained as purely serial attentive searches. Wolfe's Guided Search (GS) model accounts for this efficiency by proposing that preattentive feature processes could guide the deployment of attention in conjunction searches (Wolfe, 1993, 1994). No preattentive process can identify a particular conjunction, however, two different preattentive processes (e.g., a color and an orientation process) can cooperate to mark conjunctive targets. If the output of these two preattentive processes is combined into an attention-guiding activation map, attention will be guided to conjunction items (e.g., black vertical bars in a field of white or black horizontal or vertical bars). Wolfe's model is discussed further in Section 17.6.1.

Relatively few studies have addressed the relationship between eye movements and the search process (Findlay & Gilchrist, 1998). Findlay and Gilchrist argue that the tradition in search research to pay little attention to eye movements and instead to use the concept of covert visual attention movements (redirecting attention without moving the eyes; an important component of FIT) is misguided. The authors demonstrate that when viewers are free to move their eyes, no additional covert attentional scanning occurs. They show that unless instructions explicitly prevent eye movements, subjects in a search task show a natural propensity to move their eyes, even in situations where it would be more efficient not to do so. The authors suggest that the reason for this preference is that in naturally occurring search situations, eye movements form the most effective way of sampling the visual field.

Findlay (1997) recorded eye movements during tasks of a simple feature search and a color shape feature conjunction search. Eye movements were recorded by having the subject wear a contact lens-type search coil positioned at the center of two large Helmholtz field coils. The induced currents in the eye coils measured eye position in space in a way that minimized head movement artifact, measuring eye position with an accuracy of 10 min arc or better following calibration. Using a single feature (color) search task, in the homogeneous distractor condition, only 0.5% of first saccades were directed at a nontarget and in the heterogeneous distractor condition the percentage of misdirected saccades was under 2%. This accuracy was achieved with no cost in the time needed to program the saccade. This provided an impressive confirmation that search for a prespecified color target can be carried out in parallel. When two targets are presented simultaneously in neighboring positions, the first saccade is directed toward some "center of gravity" position. Results suggest that the control of the initial eye movement during both simple and conjunction searches is through a spatially parallel process.

Results from a conjunction search experiment (color and shape) are particularly relevant because this is a situation where serial scanning would be expected according to the classical search theory of Treisman and Gelade (1980). If a rapid serial scanning with covert attention could occur before the saccade is initiated, it is not clear why incorrect saccades would occur as frequently as observed in the experiment (three subjects were able to locate targets with a single saccade on 60–70% of occasions in the inner ring of a two-ring concentric display, and 16–40% in the outer ring). Moreover, Findlay argues that the data place constraints on the speed of any hypothetical serial scanning process because it would be necessary for a number of locations to be scanned before the target is located, given the accuracy obtained. Alternative accounts of the visual search process have appeared that assign much more weight to the parallel processes and avoid the postulation of rapid serial scanning. The results of the conjunction search are consistent with a search model that limits parallel scanning to about eight items but requires serial search for displays of larger number.

Bertera and Rayner (2000) had viewers search through a randomly arranged array of alphanumeric characters (i.e., letters and digits) for the presence of a target letter. They used both the moving window technique and the moving mask technique. The number of items in the array was held constant, but the size of the display varied ($13°\times 10°$, large, $6°\times 6°$, medium, and $5°\times 3.5°$, small).

In the moving window study, search performance reached asymptote when the window was $5°$. The moving mask had a deleterious effect on search time and accuracy, indicating the importance of foveal vision during search; the larger the mask, the longer the search time. For the window conditions, six different window sizes were used. The window was $1.0°$, $2.3°$, $3.7°$, $5.0°$, or $5.7°$; in addition, a control condition was run in which the entire array was presented (i.e., there was no window). Six different mask sizes were also used: 0.3°, 1.0°, 1.7°, 2.3°, 3.0°, or no mask present. A Stanford Research Institute Dual Purkinje Eye Tracker was used to record eye movements, with five subjects participating in the study.

Under the window condition, search performance improved (i.e., search time decreased) as the window size increased. In general, search performance reached asymptote when the window was 5.0°. There was no effect of either window size or array size on accuracy; the subjects successfully located the target letter 99% of the time across the different conditions. Under the mask condition, as mask size increased, search time increased. A large mask was more detrimental to search than a small window. Unlike window size, the size of the mask had a significant effect on accuracy. The accuracy values were 100%, 99%, 94%, 73%, 58%, and 39% for mask sizes 0 (no mask), 0.3°, 1.0°, 1.7°, 2.3°, and 3.0°, respectively.

Bertera and Rayner's study shows that a useful perceptual span in visual search extends to about 5°. The moving window paradigm for scene perception and visual search is discussed further as an instance of gaze-contingent display technology in Section 20.2.

Recently, Greene and Rayner (2001) showed that familiarity with distractors around an unfamiliar target facilitates visual search. Eye movements were recorded (right eye only) by an SMI EyeLink head-mounted tracker. Eye positions were sampled at 250 Hz by an infra-red video-based system that also compensated for head movements. Gaze positions were accurate within .5°. A saccade was recorded when eye velocity exceeded 35 deg/s or eye acceleration exceeded 9500 deg/s^2. Results indicated comparably long, but fewer fixations when distractors were familiar, discounting the theory that unfamiliar distractors need longer processing.

Results from search studies begin to call into question the role of covert attention in the search process (Findlay & Gilchrist, 1998). Findlay and Gilchrist advocate that a reappraisal of the role of covert attention in vision is in order because visual search does not provide a rationale for the existence of the covert attentional mechanism. Search situations in which the use of covert attention is advantageous are artificial and unusual. Most search tasks will be served better with overt eye scanning, guiding the eye as well as possible from the information that is being processed in parallel over the central regions of the visual field. The authors felt able to reject with some confidence the possibility that a fast covert scan of attention takes place during fixations in visual search. Covert attention might play no role, with a purely parallel process leading to saccade destination selection. Under this assumption, the saccade would be directed to the point of highest salience in some hypothetical "salience map."

17.6.1 Computational Models of Visual Search

Recent advances in the research of low-level visual processes have led to the emergence of sophisticated computational models of visual search. These models, operating on a host of still (and sometimes moving) images, model the human visual system's processes from the cornea to the striate cortex. An early example of such a model was proffered by Doll et al. (1993). The model contained six modules:

1. A pattern perception module (simulates HVS from cornea to visual cortex)
2. A visual search module (takes clutter into account)
3. A detection module (calculates probability of target detection given information on fixation locations from visual search module)
4. A decision module (predicts tradeoff between detections and false alarms)
5. An output/executive module (calculates cumulative probability of target acquisition and "missed" acquisitions; false alarms to clutter objects)
6. A target tracking module accounting for luminance contrast

Doll et al.'s model at the time only handled luminance contrast; it did not contain modules to process chromatic contrast and motion.

Wolfe's Guided Search has undergone several revisions and "upgrades" (the current version is 3.0). Previous versions of GS successfully modeled a wide range of laboratory search tasks. The most recent revisions to the model were made in an effort to handle natural images (e.g., aerial photographs). The previous version of GS ignored two important factors: first, eye movements of human observers and the interplay of covert attentional movements with overt movements of the eyes; second, the anisotropic characteristics of the retina (visual processing is much more detailed at the fovea). The newest version of GS incorporates eye movements and eccentricity effects.

The simulation of Wolfe's Guided Search starts with a greyscale image as input (see Figure 17.5). The image is processed by an array of On- and Off-Center Units, approximating retinal ganglion cell response. These provide input to Pre-Attentive Feature Maps for brightness and orientation, which model the representation of the stimulus in V1 by a complex log transform. Next, the Attention Gate allows feature information from only one object at a time to reach the higher processes such as Object Recognition. The Attentional Gate is under the control of the Activation Map. The Activation Map is a winner-take-all network that converges on a winner about every 50 ms. Feedback from the identification state to the Activation Map selects or inhibits the selected item, permitting new items to eventually win access to the identification stage.

To simulate eye movements, the Saccade Generation stage creates a saccade map. In GS, the Saccade Generation module is the analog of the superior coliculus. Activity in the saccade map is a blurred version of activity in the activation map. The GS model leads to a cooperative relationship between eye movements and attentional

Fig. 17.5. Architecture of Guided Search 3.0. Adapted from Wolfe and Gancarz (1996) with permission © 1996 Kluwer Academic/Plenum Press.

deployments. The simulation produces data that mimic human data on a number of tasks and GS eye movements resemble those recorded in human subjects.

Wolfe's Guided Search appears to be restricted to monochrome (greyscale) still images. A similar attentional model has recently been proposed by Osberger and Maeder (1998). Osberger and Maeder's algorithm uses an Importance Map (IM) to predict the perceptual importance of segmented image regions. The IM is built by considering both low- and high-level visual attributes of segmented regions. Low-level factors include: contrast, size, and shape. High-level factors include the region's location, and foreground/background classification. Osberger and Maeder's algorithm is designed to be flexible to allow easy accommodation of future contributing modules to the IM, including one analyzing color and motion.

An attentional model that considers color, as well as other familiar visual attributes, has been proposed by Itti et al. (1998). Building on biologically plausible architectures of the human visual system, the model is related to Treisman's feature integration theory. The model architecture is shown in Figure 17.6. Starting with an input image, it is progressively low-pass filtered and subsampled to yield nine dyadic spatial scales. The multiscale image representation is then decomposed into a set of topographic feature maps. Each feature is computed via a set of linear "center-surround" operations akin to visual receptive fields. Different spatial locations then compete for saliency within each map, such that only locations which locally stand out from their surround can persist. All feature maps feed, in a purely bottom-up manner, into a master "Saliency Map" (SM). Feature maps are combined into three "conspicuity maps," for intensity, color, and orientation at each scale of the saliency map. The model's saliency map contains internal dynamics that generate attentional shifts. The SM feeds into a biologically plausible "winner-take-all" (WTA) network. The Focus Of Attention (FOA) is shifted to the winning region once it has been identified by the WTA network. The WTA is then reset by a combination of local and global inhibition mechanisms that allow the selection of a future region to which the FOA is shifted in turn.

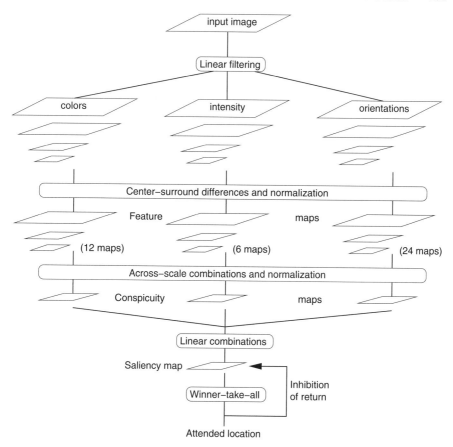

Fig. 17.6. Architecture of Itti et al.'s visual attention system. Adapted from Itti et al. (1998) with permission © 1998 IEEE.

The model has been tested on a variety of artificial and natural images, and appears to be very robust, particularly to the addition of noise. Interestingly, the model was able to reproduce human performance for a number of pop-out tasks. When a target differed from an array of surrounding distractors by its unique orientation, color, intensity, or size, it was always the first attended location, irrespective of the number of distractors. Furthermore, when the target differed from the distractors by a conjunction of features, the search time necessary to find the target increased linearly with the number of distractors. This performance is in general consistent with observations in humans and is consistent with Treisman's feature integration theory.

Of course to test any of the above visual attention models, one very attractive methodology is to compare the sequence of Regions Of Interest (ROIs) identified by an attentional model to those actually selected by human observers. A recent study by Privitera and Stark (2000) presents just such a methodology for comparing algorithmic

ROIs, or aROIs to those selected by humans, hROIs. The authors present a statistical and computational platform to perform the comparison between aROIs and hROIs. The comparison algorithm relies on two important processes: one of clustering of ROIs for comparison of loci of ROIs and a subsequent step of assembling the temporal sequences of ROIs into ordered strings of points for comparison of sequences based on *string editing*. Clustered ROIs within an image, either viewed by human observers, or selected by an attentional algorithm, are assigned character labels. Assembled strings of ROIs are compared by string editing, which, defined by an optimization algorithm, assigns unit cost to three different character operations: *deletion*, *insertion*, and *substitution*. Characters are manipulated to transform one string to another, and character manipulation costs are tabulated to yield a sequence similarity index S_s. An example is given in Figure 17.7. A positional, or loci, similarity index S_p can be found for two strings by comparing the characters of the second string to those of the first. For the strings in Figure 17.7, because all the characters of $s_2 = afbffdcdf$ are present in $s_1 = abcfeffgdc$, the two strings yield a loci similarity index of $S_p = 1$. Similarity coefficients are then sorted and stored in a table, named the Y-matrix, having as many rows and columns as the number of different sequence ROIs to be considered.

Given two strings $s_1 = abcfeffgdc$ and $s_2 = afbffdcdf$, the second can be made to equal the first by applying the following operations:

$s_1 = abcfeffgdc$
$s_2 = afbffdcdf$ *start*   *cost 0*

$s_1 = abcfeffgdc$
$s_2 = afeffdcdf$ *substitution of first b by e*   *cost 1*

$s_1 = abcfeffgdc$
$s_2 = abcfeffdcdf$ *insertion of bc after first a* *cost 2*

$s_1 = abcfeffgdc$
$s_2 = abcfeffdc$ *deletion of last df*   *cost 2*

$s_1 = abcfeffgdc$
$s_2 = abcfeffgdc$ *insertion of g*   *cost 1*

The total combined cost of deletions, insertions, and substitutions is 6. Relative to the original string length (9), the total cost yields a sequence similarity index between the two strings of $S_s = (1 - 6/9) = 0.34$.

Fig. 17.7. String editing example. From Privitera and Stark (2000) with permission © 2000 IEEE.

Using string editing similarity measures, Privitera and Stark evaluated six different attentional algorithms, some sharing similarities with the above models of Wolfe,

Osberger and Maeder, and Itti et al.. Although the algorithms tested were not the actual ones proposed by the latter group of authors, it appears that a multiresolutional strategy, such as that of Itti et al., seems to be very efficient for several classes of images. In general, although the set of algorithms picked by Privitera and Stark was only a small representative sample of many possible procedures, this set could indeed predict eye fixations.

The problem of computationally modeling human eye movements, and indeed human visual search, is far from being solved. No current model of visual search is as yet complete. Recent progress in algorithmic sophistication is encouraging. Models such as those of Wolfe, Osberger and Maeder, and Itti et al. show definite promise. Certainly the capability of tracking human eye movements has yet to play a crucial role in the corroboration of any model of visual search. Carmi and Itti (2006) recently evaluated the saliency of dynamic visual cues, noting that these cues play a dominant causal role in attracting attention. Peters and Itti (2006) suggest heuristics sensitive to dynamic events as predictors of human attention in next-generation immersive virtual environments and games.

Privitera and Stark's approach to the comparison of human and algorithmic scanpaths is one of the first methods to appear to quantitatively measure not only the loci of ROIs but also the order of ROIs. Undoubtedly future evaluation of human and artificial scanpaths will play a critical role in the investigation of visual search.

Indeed, scanpath comparison has recently received a good deal of attention. New algorithms are quickly being developed and applied with interesting insights. For example, West et al.'s (2006) *eyePatterns* was recently announced, a scanpath comparison technique inspired by bioinformatics techniques used for DNA sequence comparison. West et al.'s approach is similar to Privitera and Stark's string editing, but it makes use of the Needleman–Wunsch string similarity metric rather than Levenshtein's string distance metric. The underlying dynamic programming methods are duals of each other, with Needleman–Wunsch providing significant flexibility through its use of a user-defined similarity scoring matrix (Waterman, 1989). Guan et al. (2006) relied on Needleman–Wunsch string editing to validate the Retrospective Think-Aloud (RTA) usability testing protocol. By comparing verbalized and fixated Areas Of Interest (AOIs), they found stimulated RTA to be valid and reliable (stimulated RTA refers to retrospection cued by playback of recorded eye movements, a technique particularly suitable for gaze tracked usability testing; see Chapter 20).

17.7 Natural Tasks

A host of useful factual information has been derived through psychophysical testing (e.g., spatial acuity, contrast sensitivity function, etc.). These types of studies often rely on the display of basic stimuli, e.g., sine wave gratings, horizontal and vertical bars, etc. Although certainly central to the development of such theories as feature integration, one criticism of these artificial stimuli is their simplicity. As discussed above, studies of visual search are expanding to consider more complex stimuli such as natural scenery (Hughes et al., 1996). However, viewing of pictures projected on a laboratory display still constitutes something of an artificial task. Recent advancements in wearable and virtual displays now allow collection of eye movements in more natural situations, usually involving the use of generally unconstrained eye, head, and hand movements.

Important work in this area has been reported by Land et al. (1999) and Land and Hayhoe (2001). The aim of the first study was to determine the pattern of fixations during the performance of a well-learned task in a natural setting (making tea), and to classify the types of monitoring action that the eyes perform. Results of this study indicate that even automated routine activities require a surprising level of continuous monitoring. A head-mounted eye-movement video camera was used, which provided a continuous view of the scene ahead, with a dot indicating foveal direction with an accuracy of about 1 deg. Foveal direction was always close to the object being manipulated, and very few fixations were irrelevant to the task. Roughly a third of all fixations on objects could be definitely identified with one of four monitoring functions: locating objects used later in the process, directing the hand or object in the hand to a new location, guiding the approach of one object to another (e.g., kettle and lid), and checking the state of some variable (e.g., water level). Land et al. (1999) conclude that although the actions of tea-making are "automated" and proceed with little conscious involvement, the eyes closely monitor every step of the process. This type of unconscious attention must be a common phenomenon in everyday life.

Investigating a similar natural task, Land and Hayhoe (2001) examined the relations of eye and hand movements in extended food preparation tasks. The paper compares the task of tea-making against the task of making peanut butter and jelly sandwiches. In both cases the location of foveal gaze was monitored continuously using a head-mounted eye tracker with an accuracy of about 1 deg, and the head was free to move. In the tea-making study the three subjects had to move about the room to locate the objects required for the task; in the sandwich-making task the seven subjects were seated in one place, in front of a table. The eyes usually reached the next object in the sequence before any sign of manipulative action, indicating that eye movements are planned into the motor pattern and lead each action. The eyes usually fixated the same object throughout the action upon it, although they often moved on to the next object in the sequence before completion of the preceding action. Specific roles of individual fixations were found to be similar to roles in the tea-making task (see above). Land and Hayhoe argue that, at the beginning of each action, the oculomotor system is supplied with the identity of the required objects, information about its location, and instructions about the nature of the monitoring required during the action. The eye movements during this kind of task are nearly all to task-relevant objects, and thus their control is seen primarily top-down, and influenced very little by the "intrinsic salience" of objects. General conclusions provided by Land and

Hayhoe are that the eyes provide information on an "as needed" basis, but that the relevant eye movements usually precede the motor acts they mediate by a fraction of a second. Eye movements are thus in the vanguard of each action plan, and are not simply responses to circumstances. Land and Hayhoe conclude that their studies lend no support to the idea that the visual system builds up a detailed model of the surroundings and operates from that. Most information is obtained from the scene as it is needed.

A good deal of additional work on eye movement measurement during natural tasks has also been performed by a group of researchers seemingly co-located around Rochester, NY, mainly at the University of Rochester and the Rochester Institute of Technology. This group of researchers has been investigating the relationship among eye, head, and hand movements for some time (among other topics).

Ballard et al. (1995) investigated the use of short-term memory in the course of a natural hand–eye task. The investigation focused on the minimization of subjects' use of short-term memory by employing deictic primitives through serialization of the task with eye movements (e.g., using the eyes to "point to" objects in a scene in lieu of memorizing all of the objects' positions and other properties). The authors argue that a deictic strategy in a pick-and-place task employs a more efficient use of a frame of reference centered at the fixation point, rather than a viewer-centered reference frame that might require memorization of objects in the world relative to coordinates centered at the viewer. Furthermore, deictic strategies may lead to a computational simplification of the general problem of relating internal models to objects in the world. Sequential, problem-dependent eye movements avoid the general problem of associating many models to many parts of the image simultaneously. Thus, in a pick and place task, the steps required to perform the task of picking up a certain object from a group of objects can employ deictic references by fixating the object to be picked up next, without having to internalize a geometric reference frame for the entire set of objects.

Ballard et al. tested the above deictic reference assumption to see whether humans in fact use their eye movements in a deictic fashion in the context of natural behavior. A head-mounted eye tracker was used to measure eye movements over a three-dimensional physical workplace block display, divided into three areas, the model, source, and workspace. The task assigned to subjects was to move and assemble blocks from the source region to the workspace, arranging the blocks to match the arrangement in the model area. An example of the setup is shown in Figure 17.8. By recording eye movements during the block pick-and-place task, the authors were able to show that subjects frequently directed gaze to the model pattern before arranging blocks in the workspace area. This suggests that information is acquired incrementally during the task and is not acquired in toto at the beginning of the tasks. That is, subjects appeared to use short-term memory frugally, acquiring information just prior to its use, and did not appear to memorize the entire model block configuration

Fig. 17.8. Eye tracking in a natural pick-and-place task. Courtesy of Jeff Pelz, Visual Perception Laboratory, Carlson Center for Imaging Science, Rochester Institute of Technology <http://www.cis.rit.edu/people/faculty/pelz/-research/ASL_tracker.html>. Reproduced with permission.

before making a copy of the block arrangement.

In a similar block-moving experiment, Smeets et al. (1996) were able to show that horizontal movements of gaze, head, and hand followed a coordinated pattern. A shift of gaze was generally followed by a movement of the head, which preceded the movement of the hand. This relationship is to a large extent task-dependent. In goal-directed tasks in which future points of interest are highly predictable, the authors hypothesize that although gaze and head movements may decouple, the actual position of the hand is a likely candidate for the next gaze shift.

A recent example of such intentional visual attention was demonstrated by Pelz, Canosa, and Babcock (2000). Using a wearable eye tracker, Pelz was able to show intentionally based eye movements, which he termed "lookahead" eye movements, during a simple handwashing task. In this experiment, a subject donned a wearable computer and eye tracking rig (the computer was worn in a backpack). During a simple handwashing task, recorded eye movements showed that gaze moved to a location (soap dispenser) prior to the movement of the hands to the same location (see Figure 17.9).

To further examine issues raised by observations of natural behavior, Hayhoe et al. (2002) have recently begun using complex virtual environments that can be manipulated by the experimenter at critical points during task performance. In a virtual environment where subjects copy toy models, the authors show that regularities in

Fig. 17.9. Eye tracking in a natural hand-washing task. From Pelz et al. (2000) © 2000 ACM, Inc. Reprinted by permission.

the spatial structure are used by subjects to control eye movement targeting. Other experiments in a virtual environment with haptic feedback show that even simple visual properties such as size are not continuously available or processed automatically by the visual system, but are dynamically acquired and discarded according to the momentary task demands.

17.8 Eye Movements in Other Information Processing Tasks

Eye movements have been recorded and studied in a host of information processing tasks. Rayner (1998) provides a comprehensive review of eye movement work from multiple domains. The reader is referred to Rayner's article for the full account, the remaining classes of eye tracking applications not discussed above are listed here in point form. Unless otherwise noted, the information on this research comes from Rayner (1998).

Auditory Language Processing

In this paradigm, eye movements are recorded as people listen to a story or follow instructions regarding an array at which they are looking. Cooper (1974) introduced this method and found that when people are simultaneously presented with spoken language and a visual field containing elements semantically related to the informative items of speech, they tend to spontaneously direct their line of sight to those elements that are most closely related to the meaning of the language currently heard. Cooper observed three main types of visual behavior: (1) a visual–aural interaction mode, in which fixation of targets was correlated with the meaning of concurrently heard language; (2) a free-scanning mode, in which the subject continually altered his direction of gaze in a manner independent of the meaning of concurrently heard language; (3) a point-fixation mode, in which the subject continued to fixate the same location independent of the meaning of concurrently heard language. It was frequently the case that subjects would vacillate between more than one of these modes during the presentation of a single story or comprehension test. Informal evidence based upon subjects' postexperimental verbal reports suggests that these three types of visual behaviors may be related to their distribution of attention between the visual and auditory modalities.

The eye movement paradigm has also been applied to auditory language processing by Allopenna et al. (1998). These authors used a paradigm in which participants followed spoken instructions to manipulate either real objects or pictures displayed on a computer screen while their eye movements were monitored using an Applied Science Laboratories (ASL) E4000 eye tracker which features a lightweight camera mounted on a headband. Allopenna et al. found that eye movements to objects in the workspace are closely time-locked to referring expressions in the unfolding speech stream, providing a sensitive and nondisruptive measure of spoken language comprehension during continuous speech.

Allopenna et al. (1998) addressed two important methodological issues with the eye tracking paradigm. First, they showed that the use of a restricted set of lexical possibilities does not appear to artificially inflate similarity effects. In particular, no evidence for rhyme effects was found with successive gating, which is a task that emphasizes work-initial information. Second, the authors provided clear evidence in support of a simple linking hypothesis between activation levels and the probability of fixating on a target. The predicted probability that an object would be fixated over time closely corresponded to the behavioral data. The availability of a mapping between hypothesized activation levels and fixation probabilities that can be used to generate quantitative predictions means that eye movement data can be used to test detailed predictions of explicit models.

Allopenna et al. argue that the sensitivity of the response measure coupled with a clear linking hypothesis between lexical activation and eye movements indicates that this methodology will be invaluable in exploring questions about the microstructure of lexical access during spoken word recognition. Eye movement methodology should be especially well suited to addressing questions about how fine-grained acoustic information affects word recognition. Allopenna et al. believe that a particularly exciting aspect of the methodology is that it can be naturally extended to issues of segmentation and lexical access in continuous speech under relatively natural conditions.

Mathematics, Numerical Reading, and Problem Solving

In this area of investigation, eye movements are recorded as participants solve math and physics problems, as well as analogies. Not surprisingly, more complicated aspects of the problems typically lead to more and longer fixations.

Eye Movements and Dual Tasks

This methodology involves examination of eye movements when viewers are engaged in a dual-task situation; for example, a speeded manual choice response to a tone is made in close proximity to an eye movement. Although there is some slowing of the eye movement, the dual-task situation does not yield the dual-task interference effect typically found.

Face Perception

When examining faces, people tend to fixate on the eyes, nose, mouth, and ears. Fixations tend to be longer when comparisons have to be made between two faces rather than when a single face is examined.

Illusions and Imagery

These studies deal with illusions, such as the Necker cube or ambiguous figures.

Brain Damage

Brain damage studies have examined eye movements of patients with scotomas and visual neglect as they engage in reading, visual search, and scene perception.

Dynamic Situations

Eye movements have been examined in a host of dynamic situations such as driving, basketball foul shooting, golf putting, table tennis, baseball, gymnastics, walking in uneven terrain, mental rotation, and interacting with computers. Some of these applications are covered in the next chapter. Studies in which eye–hand coordination is important, such as playing video games, have revealed orderly sequences in which people coordinate looking and action.

17.9 Summary and Further Reading

This chapter presented eye tracking-related work mainly related to neuropsychology. Neuroscientific investigation linking functional brain mapping and eye tracking may seem a touch futuristic, however, functional brain imaging devices combined with eye trackers are already available. Currently such devices are undoubtedly expensive, however, it is certain that in due time these devices will be used to investigate visual attentional phenomena from entirely new perspectives.

The chapter focused mainly on traditional psychological investigations of vision featuring eye tracking: vision during reading, visual search, perception of art, and vision in natural and virtual environments. As can be seen, there is a good deal of potential for collaboration among computer scientists, eye tracking researchers, and scientists investigating visual perception. Providing the capability of recording eye movements over new and complex stimuli will undoubtedly extend our knowledge of vision and visual perception.

Good sources of information for keeping track of this type of work are scientific journals and conferences dealing with vision, psychology, and eye tracking. Examples include *Vision Research, Behavior Research Methods, Instruments, and Computers (BRMIC)*, and the proceedings of the European Conference on Eye Movements (ECEM), and the U.S.-based Eye Tracking Research & Applications (ETRA).

Industrial Engineering and Human Factors

Eye tracking offers a unique measure of human attentional behavior. This is particularly important in evaluating present and future environments in which humans do and will work. The examination of human interaction and behavior within their environments, particularly ones in which humans often perform functions critical to safety, is one topic studied by human factors and industrial engineers. Traditional measurement methods of human performance often include measures of reaction time and accuracy; e.g., how fast a person completes a task and how well this task is performed. These are generally measures associated with performance. To study the steps taken to perform the tasks requires analysis of the individual procedures performed. For this analysis, process measures are often needed. Eye movements are particularly interesting in this latter context because they present measures that can provide insights into the visual, cognitive, and attentional aspects of human performance.

Here, three broad experimental domains are presented in which eye tracking can play an important analytical role: aviation, driving, and visual inspection. With the development of sophisticated simulators for this task, and incorporation of eye trackers into these simulators, analysis of eye movements provides a powerful additional mechanism for measuring human factors.

18.1 Aviation

Since their early use in flight simulators (e.g., see Kocian (1987) and Longridge et al. (1989)), eye trackers have continued their utility in aviation experiments, ranging from testing procedural training to the evaluation of increasingly sophisticated deployment of new (graphical) displays. With improved simulator and eye tracking technology, the use of gaze recording techniques will undoubtedly continue.

An example of a recent combined use of relatively new eye tracking technology in a sophisticated flight simulator was reported by Anders (2001). Eye and head movements of professional pilots were recorded under realistic flight conditions in an investigation of human–machine interaction behavior relevant to information selection and management as well as situation and mode awareness in a modern glass cockpit. The simulator used was the Airbus 330 full flight simulator certified for airline training. Microphones and video cameras inside the simulator recorded environmental sounds, voice communication, general cockpit settings, and the actions of the operators. Several flight trials were conducted, with each flight starting from level 210 (21,000 feet altitude) in managed descent mode including an ILS (Instrument Landing System) approach and landing. The captain's eye and head movements were recorded during trials for later Point Of Regard (POR) analysis within a previously defined 3D model of the environment. Areas of interest were chosen to augment eye movement analysis. Major displays and panels were selected, including Primary Flight Display (PFD), Navigation Display (ND), Engine and Warning Display (EWD), System Display (SD), Main Panel (MPn), Glare Shield (GSc), and Overhead + Central Pedestal (OCA). The Flight Control Unit (FCU) is part of the GSc whereas the two Multipurpose Control and Display Units (MCDU) in front were separately defined as CDU. MAP described the chart and other paper information, with OUT representing the window area. The predefined cockpit regions of interest are shown in Figure 18.1.

Fig. 18.1. A330 cockpit with predefined areas of interest. Courtesy of Geerd Anders.

Analysis of eye movements illustrates the importance of the PFD as the primary source of information during flight (the PFD is the familiar combined artificial horizon and altimeter display gauges for combined altitude and attitude awareness). Within the PFD, most of the fixation time is attributed to flight parameters presented on the speed band (SPD), artificial horizon (AH), and altitude band (ALT). Cumulative fixation time on the heading band (HDG) is fairly low because the ND gives a better indication of horizontal situation. Average attention allocation to areas of interest is very similar for all tested approach flights. Between-subject differences (across pilots) are reportedly not outstanding. Attention shifts from the PFD to OUT shortly before landing. Attentional shifts were also recorded to the GSc in response to simulated Air Traffic Control (ATC) instructions to change flight parameters. Attention is high to the MAP and CDU areas during phases of flight in which the pilot is briefing the approach or when the flight plan changes (e.g., after a "Go direct..." or runway change instructions from ATC). As a proof of concept, this study shows the potential of eye movements for judgment of pilots' performance and future training of novice pilots.

Besides general scanpath patterns in the cockpit, eye movements have also been used to evaluate the usability of specific instruments such as newly developed electronic maps. Electronic maps such as 3D graphical navigational displays are becoming increasingly viable alternatives to traditional maps in the cockpit (Ottati et al., 1999). In their study, Ottati et al. compared eye movement patterns on different terrain features between experienced and novice pilots during a Visual Flight Rules (VFR) simulation. Based on previous studies of pilots' eye movements during Instrument Flight Rules (IFR) navigation, the authors expected experienced aviators to spend less time finding and fixating on their navigational landmarks, whereas novices were expected to have greater difficulty finding landmarks and extracting useful data from them, causing greater dwell times. As expected, a greater tendency in novice pilots was found to "fly out of the window" (i.e., devoting visual attention outside the cockpit) than in experienced pilots. Independent of expertise, pilots spent more time out of the window than over instruments.

The primary difference between experienced and novice pilots' eye movements during VFR flight reported by Ottati et al. was the amount of significant information-gathering fixations exhibited by each group. Experienced pilots employed significantly more fixations, yet each fixation lasted for relatively the same duration. This may indicate that novices were ineffectively scanning their forward field of view (outside the window) to gain sufficient navigational information. Experienced pilots' performance suggests that their fixations out of the window were more deliberate and informative. The authors suggest that if novices were trained in experienced pilot scan strategies, the evident performance gap between novices and experts could possibly be reduced.

In a study of electronic maps for taxiing, Graeber and Andre (1999) also suggest that training is necessary to assure proper usage of, and optimal visual attention interaction with Electronic Moving Maps (EMMs). The EMM studied by the authors is a display developed by NASA to increase navigation situation awareness, decrease navigation errors, increase forward taxi speeds, decrease planning time, decrease navigation mental workload, and improve navigation communications of pilots in low-visibility conditions. The main objective of Graeber and Andre's study was to understand how pilots visually interact with the EMM. This objective was partially motivated by concerns of EMMs disproportionately drawing the pilot's eyes into the cockpit during taxi, because taxiing is a primarily "eyes-out" task. Pilots navigate the airport and terminal areas using visual cues and signage from the outside environment, and through communication with ground control. Their task is to accurately navigate the cleared route, monitor for potential incursions, and maintain a safe distance from other aircraft, ground vehicles, and obstacles, a task best served by keeping their heads up and eyes out the cockpit windshield.

To examine pilots' visual interaction with the EMM and outside environment, Graeber and Andre examined pilots' eye movements under three different visibility conditions: daytime high visibility (Day Visual Meteorological Conditions, VMC), daytime low visibility (Day 700 Runway Visual Range, or RVR), and nighttime moderate visibility (Night 1400 RVR). Results indicate that visibility significantly affected the amount of time pilots dwelled on the EMM, with dwell time significantly higher in Day VMC than in Day 700 RVR (i.e., under high visibility conditions). That is, as visibility degrades, pilots spend more time eyes-out and less time dwelling on the EMM with no loss in taxi performance. A potential explanation for this surprising result is that pilots need to be eyes-out to maintain lateral and directional loop closure, scan for hazards, and maintain information gathering Out The Window (OTW). This explanation is supported by the increased percent time on route found in the performance data as the visibility decreased. As intended in the design of the EMM, it appears that pilots used the EMM as a secondary navigation aid across all visibility conditions. That is, when visibility degrades, the EMM appears to benefit performance the most, but without incurring additional visual attention requirements.

18.2 Driving

Another type of simulator in which eye trackers are increasingly being incorporated is the driving simulator. It is widely accepted that deficiencies in visual attention are responsible for a large proportion of road traffic accidents (Chapman & Underwood, 1998). An understanding of the visual search strategies of drivers is thus extremely important, and much research has been conducted in this area. Several important driving studies incorporating an eye tracker are described in Underwood (1998).

Eye movement recording and analysis provide important techniques for understanding the nature of the driving task and are important for developing driver training strategies and accident countermeasures (Chapman & Underwood, 1998). Chapman and Underwood indicate that at their simplest, drivers' fixation patterns on straight roads can be described as concentrating on a point near to the focus of expansion (the point in the visual field in front of the driver where objects appear stationary) with occasional excursions to items of road furniture and road edge markers. This reliance on the focus of expansion in the scene is assumed to be because it provides precise directional information to the driver and is the location near to which future traffic hazards are likely to be first visible. Increasing the complexity of the visual scene (by adding vehicles, road furniture, or irrelevant signing) increases the number of eye movements made and decreases the mean fixation durations on individual objects. This seems to be a natural response to having more objects available in the visual field to look at; it is not clear whether decreases in fixation durations mean that objects are processed less completely or that redundant fixation time is simply reduced.

Chapman and Underwood's study reports on recorded eye movements of relatively large groups of both novice and experienced drivers while watching videos of dangerous situations. The largest overall difference between the groups was that novices had generally longer fixation durations than experienced drivers in this task. Chapman and Underwood argue that this reflects the additional time required by novices to process information in the visual scene. Important differences were also found between the types of situation used in the study. Rural situations, even those chosen to be particularly dangerous, generally evoked fewer responses from subjects and longer fixation durations. In urban films both groups of subjects reported many more dangerous events and had shorter mean fixation durations. It is clear that dangerous events generally evoke long fixation durations. This result demonstrates the dangers in averaging eye movement data over periods of time in dynamic scenes. It is therefore argued that to fully understand the subtlety of such data, and to draw realistic conclusions about the cognitive process underlying observable behaviors, it is necessary to develop a detailed understanding of the moment-by-moment "syntax" of driving situations.

Dishart and Land (1998) discuss previous work showing that experienced drivers obtain visual information from two sections of their view of the road ahead, in order to maintain a correct position in lane while steering their vehicle around a curve. The more distant of these two sections is used to predict the road's future curvature. This section is optimally 0.75–1.00 s ahead of the driver and contains the tangent point. This section of road is used by a feedforward (anticipatory) mechanism that allows the driver to match the curvature of the road ahead. The other, nearer, section is about 0.5 s ahead of the driver and is used by a feedback (reactive) mechanism to "fine tune" the driver's position in lane. As either lane edge approaches the vehicle, the driver steers away from it, correcting his or her road position. Experiments using video-based eye–head tracking equipment have shown that the feedback mechanism is present in most people regardless of their driving experience (although its accuracy is higher in those with experience), but that the feedforward mechanism is learned through experience of steering tasks (that can include riding a bicycle, computer driving games, etc.).

Dishart and Land (1998) discuss eye–head tracking experiments on learner drivers during their tuition which indicate that use of the section of the road containing the tangent point increases with experience, then decreases as drivers learn to optimize their visual search patterns. This affords experienced drivers to spend more of their visual resources on other visual tasks both related and unrelated to driving.

A fundamental problem in visual search in driving research is defining and controlling demand on the driver as an independent variable (Crundall et al., 1998). Possible confounding factors are increases in visual demand, such as an increase in visual clutter or complexity, and increases in cognitive demand, such as an increase in the processing demands of a particular stimulus perhaps due to an increase in its relevance to a current context. Despite the lack of consistency in the manipulations of visual demand, it seems fairly well documented that general increases in task demands and visual complexity tend to reduce mean fixation duration and increase the sampling rate. Crundall et al. discuss the effects of cognitive and visual demand upon visual search strategies during driving, and whether they can be used to differentiate between novices who have recently passed their test and more experienced drivers. The authors review previous studies with the aim of identifying a process that may account for the effects of changes in demand according to driver experience.

In a driving study examining the effects of clutter, luminance, and aging, but not quite in a driving simulation, Ho et al. (2001) recorded subjects' visual search for traffic signs embedded in digitized images of driving scenes. Example driving stimulus images are shown in Figure 18.2. In a visual search task, a traffic sign was first presented to subjects, followed by a scene in which subjects were instructed to search for the sign. Analysis of five dependent measures are reported: (a) errors, (b) reaction time, (c) fixation number, (d) average fixation duration, and (e) last fixation duration. For each measure the effects of age, clutter, luminance, and target presence were evaluated. On average, older adults were less accurate than younger ones. Accuracy for daytime scenes was independent of target presence, but errors for nighttime scenes were more common on target-present trials than on target-absent trials. In daytime scenes errors were generally more common in high clutter than in low clutter, whereas in nighttime scenes no consistent clutter effect was detected. Reaction time data indicated that younger adults generally responded more quickly than older adults. Reaction times were also dependent on the amount of clutter and the presence of the target. Fixation number data, a measure that is strongly correlated with reaction time, indicated that older adults made more fixations than did younger adults, and more fixations were needed for high clutter and target-absent scenes than for low-clutter and target-present scenes. Nighttime scenes with high-clutter also required more fixations than did daytime scenes. Examining average fixation duration, it was found that younger adults had shorter fixation durations than did older adults, and low-clutter scenes resulted in shorter fixations than did high-clutter scenes. Relative to daytime scenes, high-clutter nighttime scenes containing a target resulted in longer average fixation durations. The last fixation duration reflects the comparison of the last fixated object with the target representation and the terminal decision regarding target presence. Age differences in this last stage of processing might accrue because of difficulties with the comparison process itself or because the elderly are more cautious in making overt responses. These data showed that age differences were greater on target-present trials than on target-absent trials. In summary:

- High-clutter scenes required longer latencies and more fixations to acquire the sign, were associated with more errors, and had longer fixation durations.
- Luminance effects were negligible, although high-clutter nighttime scenes did impair search efficiency on reaction time and fixation number measures. In addition, search was impaired on target-absent trials for both age groups.
- Older adults were found to be less accurate and slower than younger adults and executed more eye movements to acquire signs. The age effects on reaction time and fixation number were more pronounced on target-absent trials. Older adults used the visual cues that discriminated targets and distractors to quickly isolate the target on target-present trials. However, they took longer to correctly decide that a target sign was not present on target-absent trials, and they also took longer to make a terminal decision regarding the sign's match to the trial target on target-present trials. Interestingly, older adults were not more adversely affected by increased clutter than were the young.

Considering the rapidly aging driving population, the authors recommend that roadway engineers should consider reducing the number of competing signs (e.g., advertisements), avoiding redundant signs, and making traffic signs that are central to the safety of the driver more conspicuous.

Experiments where gaze is monitored in a simulated driving environment demonstrate that visibility of task-relevant information depends critically on active search initiated by the observer according to an internally generated schedule, which depends on learned regularities in the environment (Hayhoe et al., 2002). The driving simulator used by Hayhoe et al. consists of a steering station mounted on top of a 6 Degree Of Freedom (6 DOF) hydraulic platform. Steering, accelerator, and brake are instrumented with low-noise potentiometers that modulate signals read by a special-purpose 12-bit A/D board on an SGI VME bus. Subjects drive in "PerformerTown", an extensible environment designed by Silicon Graphics. Cars and trucks have been added to the environment that move along predefined or interactively controlled paths. The visual environment is easily modified to simulate lighting at any time of the day or night, along with the addition of fog. Observers view the display in a Head-Mounted Display (HMD) and eye position is monitored using a video-based corneal reflection eye tracker. This simulator has been used to examine "change blindness", a phenomenon where observers are insensitive to many changes made in scenes either during a saccadic eye movement or some other masking stimulus. In one experiment, while subjects drove along a prespecified path following a lead car, in a particular block a No Parking sign changed into a Stop sign for about 1 second as the observer approached it. The retinal transient caused by the change was

(a) Daytime.

(b) Simulated nighttime.

Fig. 18.2. High-clutter driving stimulus images. Reprinted with permission from *Human Factors*, Vol. 43, No. 3, 2001. Copyright 2001 by the Human Factors and Ergonomics Society. All rights reserved.

masked by 100 ms blank screens before and after the change. Results indicate that the likelihood of fixating the sign at any point while it is there is heavily modulated by the task and by the location of the sign. Given a "follow" instruction, observers rarely fixate the sign. When driving normally they invariably respond when it is at an intersection, but miss it two thirds of the time when it is in the middle of the block. This suggests observers must actively initiate a search procedure in order to see particular stimuli in the scene. Markedly different fixation patterns were displayed in the two instructions. In the "drive normally" conditions, subjects spent much more time fixating in the general region of the intersection. This strongly suggests fixation patterns and attentional control in normal vision is learned. Furthermore, the visibility of traffic signs depends on active search according to an internally generated schedule, and this schedule depends both on the observer's goals and on learned probabilities about the environment. The experiment suggests that a fuller understanding of the mechanisms of attention, and how attention is distributed in a scene, needs to be situated in the observer's natural behavior.

Research on the effects of mental activity during driving suggests the convenience of raising drivers' awareness about the possible consequences of driving while their attention is focused on their own thoughts, unrelated to driving (Recarte & Nunes, 2000). Recarte and Nunes studied the consequences of performing verbal and spatial-imagery tasks on visual search when driving. On each of four routes (two highways and two roads), participants performed two verbal tasks and two spatial-imagery tasks while their eye movements were recorded. The same results were repeated on all routes. Pupillary dilation indicated similar effort for each task. Visual functional-field size decreased horizontally and vertically, particularly for spatial-imagery tasks. Compared with ordinary driving, fixations were longer during the spatial-imagery task. With regard to driving performance, glance frequency at mirrors and speedometer decreased during the spatial-imagery tasks. Performing mental tasks while driving caused an increased attentional workload on ordinary thought, as shown by pupillary dilation. Mental tasks imposed by the experimenter while the participant was driving, and, therefore, of a more mandatory nature than ordinary thoughts, produced: (a) marked changes in the visual inspection patterns; (b) qualitatively different changes, depending on the type of processing resources required by the mental tasks; (c) the same effects in the four different driving scenarios; and (d) changes in practical driving behaviors, such as inspection reduction of mirrors and speedometer. Spatial-imagery tasks produced more marked effects in almost all the analyzed variables than did verbal tasks. First, an increment in mean fixation durations was found, due to some long fixations and possibly associated with mental image inspection as part of the mental activity of image search or rotating. It is suggested that these eye freezing responses produce impairment of environment perception. Second, a marked reduction of the visual inspection window was found, both horizontally and vertically, possibly associated with narrowing of the attentional focus size. Smaller saccadic size and marked reduction of glance frequency at mirrors and speedometer were also noted. With regard to the implications for driving, Recarte and Nunes suggest that the spatial reduction of the visual inspection window, including the reduction of the inspection of mirrors, could be interpreted as a predictor of decreased probability of detecting traffic events, particularly when performing mental spatial-imagery tasks.

It has been experimentally demonstrated that the pattern of eye fixations reflects, to some degree, the cognitive state of the observer (Liu, 1998). Liu suggests that the analysis of drivers' eye movement may provide useful information for an intelligent vehicle system that can recognize or predict the driver's intention to perform a given action. Such a system could improve the interaction between the driver and future vehicle systems and possibly reduce accident risk. Liu reviews numerous studies that have tried to establish the level at which the relationship between eye movements and higher cognitive processes can be modeled. The least controversial conclusions simply state that the pattern of eye movements generally reflects the observer's thought processes, indicating to some degree the goals of the observer and perhaps even the main areas of interest. The strongest conclusions assert that the eye movements are directly observable indicators of underlying cognitive processes, revealing the nature of the acquired information as well as the computation processes. To implement a "smart car," one which would be able to predict or recognize the driver's intentions and then take the appropriate course of action based on that prediction, Liu suggests that it may be possible to determine the current driver state (e.g., centering of the car in the current lane, checking whether the adjacent lane is clear, steering to initiate a change in heading, centering the car in the new lane, etc.) from observed eye movement patterns via hidden Markov dynamic models (HMDMs). The primary consideration for including driver eye movements in HMDMs is how the eye movement behavior will be coded (e.g., by gaze locations or by gaze or saccade direction). Gaze location seems to be the natural choice given that characteristic patterns of driver eye movements have been identified using a Markovian analysis of gaze location. The results from preliminary experiments using HMDMs without utilizing eye movements are promising. The addition of eye movement analysis should enhance the system performance by enabling recognition of driver intentions rather than just recognition of maneuvers as they begin.

18.3 Visual Inspection

In a study of visual inspection of integrated circuit chips Schoonard et al. (1973), state that "visual inspection pervades the lives of all people today. From poultry, meat, and fish inspection, to drug inspection, to medical X-ray inspection, to production line inspection, to photo interpretation, the consequences of inspection directly affect people's lives through their effects on the quality and performance of goods and services."

Important criteria of evaluating the efficiency of human visual inspection include measurements of performance, as indicated by the inspector's speed and accuracy. These performance measures effectively summarize general outcomes of the process of (visual) inspection. In contrast, eye movements captured during visual inspection provide visualization of the inspector's process, and therefore provide an instance of process measures. Important eye movement-related process measures include fixation durations, number of fixations, interfixation distances, distribution of fixations, fixational sequential indices (order of fixations), and direction of fixations. The utility of process measures relies on their relation to, or explanation of, performance. Thus one expects to find a relation between process and performance measures.

With respect to performance, a key issue is the ability to gauge the visual search process with respect to some identifiable baseline measurement. For example, for training or evaluation purposes, it is desirable to be able to compare an inspector's visual search process to that of an expert inspector, or to some ideal model of performance. Visual search can be described by a two-stage process:

1. A time-consuming visual search for a target
2. A decision process identifying the found item as either target or nontarget

Visual search can be modeled by two idealized processes: a perfect-memory systematic (serial) search, or a memoryless random search. A systematic search differs from a random search principally in the manner of visitation of searchable regions. That is, systematic search, due to its perfect memory, will visit each portion of the entire search region only once. In contrast, under random search, every portion of the search area has equal probability of being visited at any time during the search. Inasmuch as there is no memory of past visitations, regions may be visited more than once (re-inspected).

To model random visual search more formally, consider a visual search area A, with randomly distributed search targets (symbolized by $\times$), and a visual lobe (instantaneous fixation size) with area a, as shown in Figure 18.3. Let the probability of detecting a target (e.g., a defect) given a fixation at a particular location be s; i.e.,

$$p(\text{detecting defect} \mid \text{fixation}) = s, \ s \in (0,1).$$

Then let the probability that a fixation contains (or lands on) a target be $n_d * (a/A)$; i.e.,

$$p(\text{fixation contains a defect}) = n_d * \frac{a}{A},$$

where n_d is number of randomly distributed defects. Thus, the probability of detecting a defect is:

$$p(\text{detecting a defect}) = n_d * \frac{a}{A} * s = Q$$

on any single fixation. The probability of detecting a defect on any ith fixation, in a series of fixations, is:

$$p(\text{detecting a defect on } i\text{th fixation}) = (1 - Q)^{i-1}Q,$$

where $(1 - Q)$ is the probability of not detecting the fault on previous fixations. Thus the cumulative probability of detecting a defect over n fixations is:

$$Q \sum_{k=1}^{n} (1-Q)^{k-1} = [1-(1-Q)^n] = 1 - e^{-Qn}$$

$$= 1 - e^{[n_d*(a/A)*s*n]}.$$

Considering the time needed to search t, and the time devoted per fixation t', letting $n = t/t'$, the cumulative probability can be rewritten as

$$Q = 1 - e^{\left[n_d*(a/A)*s*t/t'\right]}$$

$$= 1 - e^{-\lambda t}, \text{ with } \lambda = n_d * \frac{a}{A} * \frac{s}{t'}.$$

Q is an exponential distribution that approaches probability 1 as search time increases. A stylized plot of the exponential distribution is shown in Figure 18.4 (lowest solid curve). The model essentially predicts that with more time, eventually all defects (or almost all defects) will be detected. An important attractive feature of this model is that it predicts the accuracy of detection during the first few seconds of search, meaning that the probability of detection will not change much later on in the search. In other words, an inspector will either see or not see the defect fairly quickly, but the probability of detection will not change much with increased search time.

To illustrate the predictive property of the model, consider three different search durations: $t_1 = 3s$, $t_2 = 12s$, and $t_3 = 10s$. The mean search time is approximately $25/3 = 8 \sim 10s$. The cumulative probability function for this search can be modeled by the function $1 - e^{-t/10} = 1 - e^{-.1t}$, and is plotted in Figure 18.5. This particular instance of the visual search model would predict (roughly) that about 10 seconds are needed to achieve a 63% accuracy rate. A 95% success rate is reached at about 30 seconds.

The goal of modeling visual search in a visual inspection task is to be able to gauge an inspector's performance versus an idealized model of visual search such as the random search derived above. In contrast to this memoryless model, a perfect-memory systematic search is expected to reach a higher accuracy level faster than the random model. A systematic search model may be described by a piecewise-linear function as shown in Figure 18.4 (highest solid curve). The systematic search is simply another idealized model, one that happens to be very efficient in comparison with the random model. Human performance is expected to fall somewhere in between the two idealized models. That is, when a cumulative distribution of human visual search performance is plotted, it should appear somewhere between the two curves, as indicated by the dashed curve in Figure 18.4. Using this methodology, it is then possible to gauge the efficiency of an individual inspector: the closer the inspector's curve is to the idealized systematic model representation, the better the inspector's performance. This information can then be used to either rate the inspector, or perhaps to train the inspector to improve her or his performance.

Fig. 18.3. Model of visual search area, visual lobe, and targets.

Tracking eye movements during visual inspection may lead to similar predictive analyses, if certain recurring patterns or statistics can be found in collected scanpaths. For example, an expert inspector's eye movements may clearly exhibit a systematic pattern. If so, then this pattern may be one that can be used to train novice inspectors. In the study of visual inspection of integrated circuit chips, Schoonard et al. (1973) found that good inspectors are characterized by relatively high accuracy and relatively high speed, and make many brief eye fixations (as opposed to fewer longer ones) during the time they have to inspect.

In a survey of eye movements in industrial inspection, Megaw and Richardson (1979), identify the following relevant eye movement parameters.

- Fixation times. The authors refer to the importance of recording mean fixation times (perhaps fixation durations would be more appropriate). Megaw and Richardson state that for most inspection tasks, the expected average fixation duration is about 300 ms. Longer fixation times are associated with the confirmation of the presence of a potential target and with tasks where the search times are short.

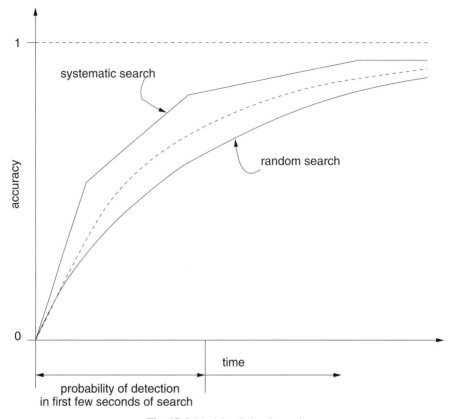

Fig. 18.4. Models of visual search.

- Number of fixations. The number of fixations is a much more critical parameter in determining search times than fixation times and is sensitive to both task and individual variables.
- Spatial distribution of fixations. The coverage given to the stimulus material can be found by measuring the frequency of fixations falling in the elements of a grid superimposed over the display or by finding the frequency of fixations falling on specific features of the display. In both cases these frequencies correlate with the informativeness of the respective parts of the display as revealed by subjective estimates made by the searchers.
- Interfixation distances. With static displays this measure is equivalent to the amplitude of the saccadic movements. It is possible that when a comparatively systematic strategy is being employed, interfixation distances may reflect the size of the useful field of view (visual lobe).
- Direction of eye movements. Horizontal saccades may occur more frequently than vertical ones, which may reflect the elliptical shape of the effective useful field of view (visual lobe).

Fig. 18.5. Example of visual search model.

- Sequential indices. The most popular of these is the scanpath.

In their survey, the authors review previous inspection studies where eye movements were recorded. These include inspection of sheet metal, empty bottles, integrated circuits, and tapered roller bearings. In an inspection study of tin-plated cans, the authors report that experienced inspectors exhibited smaller numbers of fixations and that each inspector used the same basic scanpath from one trial to the next although there were small differences between inspectors. In an inspection study of electrical edge connectors, it was noted that the number of fixations per connector was much greater for complex items (i.e., more fixations are needed to search over a complex target).

Besides being useful for gauging inspection performance, eye movements may play a part in training visual search strategy (Wang et al., 1997). Visual search strategy training can be effective in adoption of a desirable systematic search strategy. To train visual search strategy, eye movements may be used as both a feedback mechanism and as confirmation of adoption of the new search strategy. In their paper on search strategy training, Wang et al. recorded eye movements over visual inspection of artificial printed circuit boards. Eye movements were used to check how well each subject's search pattern followed the strategy trained. Scanpaths were judged by the experimenter after each training trial and feedback was given to the subject regarding whether she or he had searched in a random or systematic manner.

Because training has been identified as the primary intervention strategy in improving inspection performance (see Gramopadhye et al. (1998)), Duchowski et al. (2001) investigated the utility of eye movements for search training in virtual reality. A three-dimensional aircraft inspection simulator was developed at Clemson University for this puropose. The aesthetic appearance of the environment is driven by standard graphical techniques augmented by realistic texture maps of the physical environment. The user's gaze direction, as well as head position and orientation, are tracked to allow recording of the user's fixations within the environment. The diagnostic eye tracking VR system allows recording of process measures (head and eye movements) as well as performance measures (search time and success rate) during immersion in the VR aircraft inspection simulator. The VR simulator features a binocular eye tracker, built into the system's Head-Mounted Display (HMD), which allows the recording of the user's dynamic Point Of Regard (POR) within the virtual environment. A user wearing the HMD and an example of raw eye tracker output are shown in Figure 18.6. An experiment was conducted to measure the training effects of the VR aircraft inspection simulator. The objectives of the experiment included: (1) validation of performance measures used to gauge training effects, and (2) evaluation of the eye movement data as cognitive feedback for training. Assuming eye movement analysis correctly identifies fixations and the VR simulator is effective for training (i.e., a positive training effect can be measured), the number of detected fixations is expected to decrease with the adoption of an improved visual search strategy (Drury et al., 1997) (e.g., following training). The criterion task consisted of inspecting the simulated aircraft cargo bay in search of defects. Several defects can occur in a real environment situation. Three types of defects were selected to create inspection scenarios:

1. Corrosion: represented by a collection of gray and white globules on the inner walls of the aircraft cargo bay and located roughly at knee level.
2. Cracks: represented by a cut in any direction on the structural frames inside the aircraft cargo bay.
3. Damaged conduits: shown as either broken or delaminated electrical conduits in the aircraft cargo bay.

Data for performance and cognitive feedback measures were collected using search timing and eye movement information, respectively. The following performance measures were collected.

1. Search time from region presentation to fault detection
2. Incremental stop time when subjects terminated the search in a region by deciding the region does not contain faults
3. Number of faults detected (hits), recorded separately for each fault type
4. Number of faults that were not identified (misses)

Fixation analysis enabled the collection of cognitive feedback measures, which were provided to subjects during the training session. Cognitive feedback measures were based on the eye movement parameters that contribute to search strategies as defined by Megaw and Richardson (1979), including: (1) total number of fixations, (2) mean fixation duration, (3) percentage area covered, and (4) total trial time. Cognitive feedback measures were graphically displayed off-line by rendering a 3D environment identical to the aircraft cargo bay that was used during immersive trials. This display represented the scanpaths of each trial to indicate the subject's visual search progression. An example scanpath is shown in Figure 18.7.

Analysis indicates that, overall, training in the VR aircraft simulator has a positive effect on subsequent search performance in VR, although there is apparently no difference in the type of feedback given to subjects. Cognitive feedback, in the form of visualized scanpaths, does not appear to be any more effective than performance feedback. It may be that the most effective common contributor to training is the immersion in the VR environment, that is, the exposure to the given task, or at least to the simulated task. Whether the eye tracker, by providing cognitive feedback, contributes to the improvement of inspection performance is inconclusive. Users may benefit just as much from performance feedback alone. However, the eye tracker is a valuable tool for collecting process measures. Analysis of results leads to two observations. First, mean fixation times do not appear to change significantly following training. This is not surprising because eye movements are to a large extent driven by physiology (i.e., muscular and neurological functions) and cognitive skill. In this case the search task itself may not have altered cognitive load per se, rather, prior experience in the simulator may have facilitated a more efficient search. Second, the number of fixations appears to decrease following training. These results generally appear to agree with the expectation of reduced number of fixations with the adoption of an improved visual search strategy (e.g., due to learning or familiarization of the task). The implication of reduced number of fixations (without an increase in mean fixation time) suggests that in the posttraining case, subjects tend to employ a greater number of saccadic eye movements. That is, an improved visual search strategy may be one where subjects inspect the environment more quickly (perhaps due to familiarity gained through training), reducing the time required to visually rest on particular features.

Alternatively, fixation duration may be influenced by feedforward training, as follow-on studies have shown (Sadasivan et al., 2005). These more recent results suggested that given feedforward training (e.g., "You should look here"), participants significantly increased their mean fixation duration over those who did not benefit from training. Training benefit was observed in improved accuracy, leading to a classic speed–accuracy tradeoff. The trainees' slower-paced inspection strategy may have stemmed from a more deliberate target search/discrimination strategy that required more time to execute (see the relevant case study report in Chapter 15).

Recently, Reingold et al. (2002) reviewed the motivation behind using chess as an ideal task environment for the study of skilled performance. Because pioneering work showing perception and memory to be more important differentiators of expertise than the ability to think ahead in the search for good moves, chess research has been instrumental in enhancing our understanding of human expertise and in

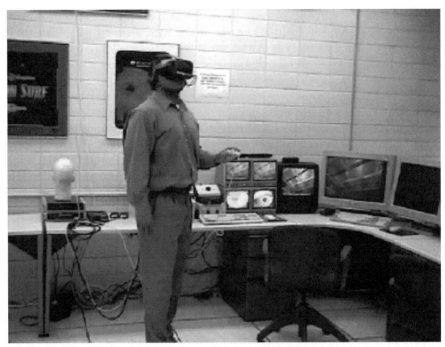

(a) User wearing eye tracking HMD.

(b) Raw eye tracker output (left eye).

Fig. 18.6. Virtual aircraft inspection simulator.

File   Scenarios   Options   Run   Help

| Cargobay | Analysis | Data Viewer | Algorithm Options |

Scenario successfully loaded

Fig. 18.7. Visualization of 3D scanpath in VR.

contributing to the study of artificial intelligence. The chess master is thought to use recognizable configurations of pieces, chunks, and templates as indices to long-term memory structures that, in association with a problem-solving context, trigger the generation of plausible moves for use by a search mechanism. Search is thereby constrained to the more promising branches in the space of possible moves from a given chess position. Hence, Grandmasters, the best human players, can find excellent moves despite generating only a small number of potential states (perhaps 100 or so for a few minutes of search). Such constrained search differs sharply from the enormous space explored by computer chess programs, which typically explore millions to hundreds of millions of alternatives in the same timeframe.

Reingold et al. study visual span as a function of chess skill (expert vs. intermediate vs. novice) and configuration type (chess configuration vs. random configuration) using a gaze-contingent window technique. The paper extends classic work demonstrating that after viewing structured, but not random, chess positions for five seconds, chess masters reproduced these positions much more accurately than lesser-skilled players. The authors document dramatically larger visual spans for experts while processing structured, but not random, chess positions. In a check detection task, where a minimized 3×3 chessboard containing a king and potentially checking pieces were displayed, experts made fewer fixations per trial, and had a greater proportion of fixations between individual pieces, rather than on pieces. These results provide strong evidence for a perceptual encoding advantage for experts attributable to chess experience, rather than to a general perceptual or memory superiority.

18.4 Summary and Further Reading

To summarize the potential impact of eye trackers in human factors research, this chapter focused on the eye tracker's capability of recording human visual process measures during exemplar tasks in aviation, driving, and during visual inspection. Performance measures typically quantify how a person performed (e.g., with what speed and accuracy), however, process measures cannot only corroborate performance gains, but can also lead to discoveries of reasons for performance improvements (i.e., what the subject performed). In particular, tracking the users' eyes can potentially lead to further insights into the underlying human cognitive processes under varying conditions and workloads.

Good sources for further information dealing with eye tracking and human factors include the Human Factors and Ergonomics Society (HFES) journal and annual conference proceedings, and the proceedings of SIGCHI, the European Conference on Eye Movements (ECEM), and the U.S.-based Eye Tracking Research & Applications (ETRA).

19

Marketing/Advertising

Eye tracking can aid in the assessment of advertising effectiveness in such applications as copy testing in print, images, video, or graphics, and in disclosure research involving perception of fine print within print media and within available television and emerging High Definition TV (HDTV) displays.

A particularly illustrative although fictitious example of eye tracking in advertising can be seen in the movie *Looker* (Crichton, 1981). In one scene, the star of the movie, Dr. Larry Roberts, a cosmetic surgeon (played by Albert Finney), is shown a potential advertisement of a beauty product. The ad features an attractive model in a beach scene. The beauty product is displayed while Dr. Roberts' eyes are tracked. When it is obvious that Dr. Roberts' attention is drawn to the attractive model rather than the beauty product, the product is immediately moved to appear much closer to the point on the screen where Dr. Roberts' attention is drawn. Although fictional, the effectiveness of eye tracking is effectively illustrated.

The motivation for utilizing an eye tracker in market research stems from the desire to understand consumer actions. In general, advertisers aim to provide product information to consumers in an efficient manner so that consumers' awareness of the existence of the product is heightened. If the consumer identifies the product as one that can potentially satisfy her or his current need, it is expected that the consumer will be more likely to purchase that particular product than if the consumer had not been aware of the product's availability. Consumer action can roughly be modeled by the block diagram shown in Figure 19.1. In general, a consumer's actions will be influenced by a combination of external and internal factors. External factors may include marketing actions (e.g., product promotion, distribution, availability), competitive factors (e.g., the desire to possess the latest and greatest product), and environmental factors (e.g., rainy conditions causing the need to possess an umbrella). Based on external influences and internal (perceptual and cognitive) processes, a consumer will make a choice as to whether to make a purchase, and which product to purchase, resulting in consumer action. The human decision-making process may be affected by recognition of one's need or desire, and may also be influenced by

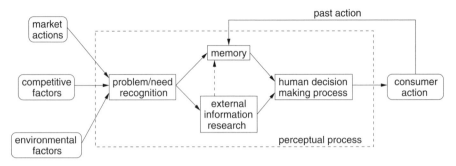

Fig. 19.1. Model of market and consumer actions.

external information gathered through research and/or through past memories and experiences.

It can be argued that external influences and resultant consumer actions are the only relevant and measurable factors to advertisers. That is, one can measure the relationship between marketing action (level of advertising) to the resulting consumer actions (sales of product). In this case, the model of the consumer may be treated as a black box: it is not as important to measure how the consumer functions, but rather just to measure what actions the consumer has performed (i.e., it may be sufficient to measure just the effect of consumer actions, not necessarily the cause). On the other hand, if one understands the cognitive and perceptual processes internal to the consumer, then a model of consumers' internal processes may aid the direction of marketing actions. A primary goal then is an understanding of the type of information that consumers want or use to make their decisions. Thus, if one can measure the perceptual process during the consumer's acquisition of information, it may be possible to tailor the information in a way such that the information the marketer wishes to impart to the consumer is delivered as efficiently and directly as possible. Eye tracking can provide insight into at least one aspect of the internal consumer model: how the consumer disperses visual attention over different forms of advertising. For example, current eye tracking technology can fairly easily provide a glimpse of the consumer's (overt) attentive processes over print media and television advertising, as illustrated above by the scene in the movie *Looker*. Indeed, it is quite plausible that market researchers are aware of eye trackers and utilize them for this purpose. Unfortunately, finding evidence of eye tracking in use by advertising companies is somewhat difficult. It appears that the use of eye trackers is not often well documented or otherwise advertised. Applied research organizations may routinely examine the eye movements of consumers as they look at advertisements, however, this work tends to be proprietary (Rayner et al., 2001). Here, only a few examples of published eye tracking work are presented, the rest of this chapter is augmented by examples of students' work and in-class projects.

19.1 Copy Testing

A particularly good example of analysis of eye movements over advertisements in the Yellow Pages™ is given by Lohse (1997). In this experiment, eye movement data are collected while consumers choose businesses from telephone directories. The study addresses (1) what particular features cause people to notice an ad, (2) whether people view ads in any particular order, and (3) how viewing time varies as a function of particular ad features. Based on a review of the literature on attention and a brief survey of prototypical eye movement patterns, the author develops the following propositions.

1. Color: users are more likely to notice color ads before any other type of ad.
2. Graphics: users are likely to notice ads with graphics before ads without graphics.
3. Size: users are likely to notice large ads before small ads.
4. Location: users are more likely to view advertisements near the beginning of the heading than those near the end of the heading.

It is interesting to note some important points of the experimental methodology. To prevent bias toward any recognizable businesses, a completely artificial mockup of a 32-page Yellow Pages directory was created. The pages were virtually indistinguishable from real Yellow Pages in terms of font, ink, and color, but all business entries were fabricated. The 32 directory pages were organized into four books to control for combinations of the following layout and design features: ad type, location of display ads on the page, size of ad, color, use of graphics, whether a listing had a bold typeface, serial position of the ad (alphabetic order), and number of types of information in the ad (hours, years in business, slogan, brand names, specialties). Eye movement analysis revealed that the results from the Yellow Pages study are consistent with previous findings on print advertising in magazines, catalogs, and newspapers. Ad size, graphics, color, and copy all influence attention to advertisements. The author offers the following observations.

1. Color and graphics: color ads with graphics captured attention. Color ads were scanned more quickly, more often, and longer than black and white ads. Subjects noticed more color ads than ads without color (92% vs. 84%) and viewed color ads before ads without color. Subjects viewed color ads 21% longer than equivalent ads without color. Subjects also viewed 96% of ads with graphics. However, unlike color, graphics did not capture initial consumer attention.
2. Size: ad size influenced attention. In general, the larger the ad, the more likely subjects were to notice the ad. Subjects noticed 93% of the large display ads but only 26% of the plain listings. Quarter-page ad displays were much more noticed than text listings.
3. Location: the position of an ad on the page had a large effect on whether people viewed the ad, even though the position says nothing informative about the business. Position matters because people scan ads on a page in alphabetic order and their scan is not exhaustive; as a result, people never read some ads.

Due to the improvement of eye tracking technology (e.g., bite bars are no longer needed), the author suggests that the time has come to reevaluate the importance of eye tracking equipment as a tool for print advertising research.

19.2 Print Advertising

In a study of consumers' visual attention over print advertisements, an eye tracker was used to gain insight into attentive processes over repeated exposure to print advertisements (Rosbergen et al., 1990). The authors explored the phenomenon of repeated advertising's "wearout"; i.e., the authors investigated consumers' diminishing attentional devotion to ads with increased repetition. Consumers' visual attention was measured to key print ad elements: headline, pictorial, bodytext, and packshot (a closeup photograph of a product). A statistical model was proposed comprising submodels for three key measures of visual attention to specific elements of the advertisement: attention onset, attention duration, and inter- and intraelement saccade frequencies. Analyses show that whereas duration decreases and attention onset accelerates during each additional exposure to the print ad, the attentional scanpath remains constant across advertising repetitions and across experimentally varied conditions. The authors also list their important findings:

- Attention durations differ significantly across ad elements. Parameter estimates reveal that attention duration is longest for the text, followed by headline, and shortest for the pictorial and the packshot. In addition, a progressive decrease in the expected attention duration is observed across exposures. Neither motivation nor argument quality affects the amount of attention paid to ad elements.

- The time until subjects first fixate on an ad element differs among elements. Subjects attend first to the headline followed by the pictorial, the text, and finally the packshot. Although repetition as such has no impact on attention onset, differences in attention onset are not constant across exposures. Less time lies between the expected starts of the first fixations during the second and third exposures than during the first exposure. In other words, the attentional process accelerates during later exposures.

    Attention onset is also significantly affected by motivation in the sense that attention onsets are farther apart for highly motivated subjects than for less motivated subjects. However, motivation does not change the order in which ad elements are attended to for the first time nor does the impact of motivation on attention onset differ across exposures.

- Analysis and modeling of scanpaths suggest that the ad's scanpath can be described by a reversible, stationary first-order Markov process. Based on this model, expected transition matrices suggest that:
    1. The amount of attention paid to the text is about three times as high as the amount paid to the pictorial.

2. The amount of attention paid to the ad decreases by about 50% from exposure 1 to exposure 3.
3. The majority of saccades (about 75%) occur within ad elements, in particular in the bodytext.
4. Most interelement saccades start from or end at the packshot.
5. The expected transition matrices are quasi-symmetric.
6. The conditional transition probabilities remain constant across exposures.

Combining the results with the authors' modeling efforts yields the following observations. As the attention onsets show, subjects attend, on average, first to the headline. As indicated by the expected number of saccades between headline and pictorial, attention is then directed to the pictorial. However, attention onsets provide some indication that during later exposures this order may be reversed. Both headline and pictorial receive about one-sixth of subjects' attention. Half of the attention is directed at the bodytext, but subjects focus on the bodytext only after the headline and the pictorial have received some initial attention. Finally, subjects attend to the packshot last, and that, despite the limited amount of attention spent on this ad element, most intraelement saccades start from and end at the packshot. This may point to integration of information in other ad elements with information in the packshot.

In a recent study of eye movements over advertisements, Wedel and Pieters (2000) comment that although eye movements are eminent indicators of attention, what is currently missing in eye movement research is a serious account of the processing that takes place to store information in long-term memory. The authors attempt to provide such an account through the development of a formal model. They model the process by which eye fixations on print advertisements lead to memory for the advertised brands. The model is calibrated to eye movement data collected during exposure of subjects to magazine ads and subsequent recognition of the brand in a perceptual memory task. Available data for each subject consist of the frequency of fixations on the ad elements (brand, pictorial, and text), and the accuracy and the latency of memory. It is assumed that the number of fixations, not their duration, is related to the amount of information a consumer extracts from an ad. The accumulation of information across multiple fixations to the ad elements in long-term memory is assumed to be additive. Wedel and Pieters's model is applied in a study involving a sample of 88 consumers who were exposed to 65 print ads appearing in their natural context in two magazines. Wedel and Pieters report that across the two magazines, fixations to the pictorial and the brand systematically promote accurate brand memory, but text fixations do not. Brand surface has a particularly prominent effect. The more information is extracted from an ad during fixations, the shorter the latency of brand memory is. A systematic recency effect was found: when subjects are exposed to an ad later, they tend to identify it better. In addition, there is a small primacy effect. The effect of the ad's location on the right or left of the page depends on the advertising context.

Considering text and pictorial information in advertisements, Rayner et al. (2001) performed a study where viewers looked at print advertisements as their eye movements were recorded. Eye movements were recorded by an EyeLink headband-mounted tracker from SensoMotoric Instruments (SMI). Although viewing was binocular, movements of the right eye were monitored. Eye positions were sampled at 250 Hz. Half the viewers were told to pay special attention to car ads, and the other half were told to pay special attention to skin-care ads. Rayner et al. found that viewers tended to spend more time looking at the text than the picture part of the ad, although they did spend more time looking at the type of ad to which they were instructed to pay attention. Fixation durations and saccade lengths were both longer on the picture part of the ad than the text, but more fixations were made on the text regions. Viewers did not alternate fixations between the text and picture part of the ad, but they tended to read the large print, then the smaller print, and then they looked at the picture (although some viewers did an initial cursory scan of the picture). Of the 110 product names that were generated in the free-recall task, 89 (81%) were recalled correctly. Overall recognition performance was excellent. Despite participants' focus on the text in the ads, memory for the product names was not particularly good. On average, participants correctly recalled fewer than four brand names from the 24 advertisements they had studied. In addition, the advertisements that were viewed favorably were often not identified by brand name; those preferred ads were often described to the experimenter in terms of some aspect of the pictorial information or by a generic product label.

Although Rayner et al. report that scan path data from the experiment are difficult to quantify, they note some very striking characteristics of how viewers scanned the ads. The initial fixation on the ad was always located in the center of the ad because that is where the viewer was fixated when the ad was initially presented. Looking behavior was fairly consistent across viewers in that they typically initially made an eye movement to the large print, regardless of its spatial position within the ad. After looking at the large print, participants either made a very cursory scan of the picture, or more typically, they moved from the large print to the small print and then to the picture.

Rayner et al. suggest that the presented data have some striking implications for applied research and advertisement development. Instructions given to participants can influence their eye movements, suggesting that participants' goals must be considered in future research conducted by advertising agencies and researchers in the area. The authors note that the data also indicate that an advertisement captures and holds participants' attention but this may be caused by the instructions given to those participants. If different instructions were used, the evidence of the advertisements' success in capturing attention may be reduced. The authors note that their second main finding, that more time was spent viewing text than viewing pictures in these ads, is inconsistent with how advertising agencies view the relative importance of the visual and text portions of the ads. Early research on print advertisements concluded that an ad is typically well liked during prepublication testing if it has a single illustration, a short headline, lots of white space and very little text. Data presented here indicate that consumers may be paying much more attention to the text in ads than previously thought. Rayner et al. conclude that currently there is no cognitive process theory that makes clear predictions about eye movements and attention while viewing print advertising. The data and ideas outlined in this paper have the potential to form the foundation in the search for such a theory.

19.3 Ad Placement

Eye movements recorded over advertisements are particularly informative because scanpaths immediately provide a visual depiction of whether the intended text or object was fixated (or at least scanned over). This was a topic of study of a pair of senior undergraduate students, one from computer science, the other from marketing. Examples of scanpaths recorded over advertisement images are shown in Figure 19.2. Although formal analyses cannot be offered, the images show fairly typical scanpaths over ads. It can be seen that attention is drawn to fairly conspicuous ad elements, such as faces, textual information, and objects set apart by virtue of being presented in homogeneous color regions.

Another marketing student study was conducted over NASCAR™ (National Association for Stock Car Auto Racing) images. The student team consisted of majors from computer science, marketing, and industrial engineering. The distribution of work fell along fairly well-expected lines of specialization: the computer science major provided coding support, the industrial engineer provided experimental design and statistical analysis expertise, and the marketing students provided contextual and background information (motivation for the study, image samples, etc.). The objective of the experiment was simply to find whether any particular regions on the vehicle tend to inherently draw attention. Images from the study are given in Figure 19.3.

Different views of NASCAR vehicles were evaluated: a front-right view (as shown in Figure 19.3), a front-left view, and left and right views of the vehicle rears. These images were chosen to represent views of vehicles as may be seen during a televised NASCAR event. Specific vehicle regions were chosen a priori as regions of interest: the hood, middle, rear quarter panel, and trunk, depending on the view of the vehicle. Pictures viewed by subjects were grouped based on the visibility of ROIs. Mean viewing times and number of fixations were compared between groups. The control stimulus was an image of a NASCAR vehicle with the ads airbrushed out (as shown in Figure 19.3(b)). The control stimulus was used as an image meant to be unbiased by advertising content. Informal analysis based on the number of fixations within ROIs suggests that the shape of the vehicle alone is responsible for drawing attention rather than the placement of the ads themselves. From images showing the left and right rear views of the vehicles, it was determined that the rear quarter panel draws most of the viewer's attention (the highest number of fixations were counted in this

(a) Altoids ad.

(b) M&Ms ad.

(c) Tommy Hilfiger ad.

(d) Hugo ad.

Fig. 19.2. Scanpaths over advertisements. Courtesy of Cristy Lander and Karen Kopp. Reproduced with permission, Clemson University.

(a) Front-right view of vehicle.   (b) Advertisements air-brushed out.

(c) Identified fixations.   (d) Highlighted Regions Of Interest (ROIs).

Fig. 19.3. Scanpaths over NASCAR™ vehicles. Courtesy of Melissa Andrews, Laura Boyd, Robyn Bushee, and Amit Joshi. Reproduced with permission, Clemson University.

ROI). From images showing the right and front views of the vehicles, the middle section was found to draw most of the viewers' attention.

Due to the short, in-class nature of these studies, significance and generalizability of results can only be claimed to be anecdotal. Results reported here are not meant to be conclusive or significant. However, the exercises are excellent examples of potential student-led eye tracking projects in testing advertisement materials.

19.4 Television Enhancements

Thematically similar to copy layout and ad placement design, Josephson and Holmes (2006) studied the addition of on-screen enhancements to TV news broadcasts. On-screen enhancements consisted of a news crawler, or ticker, usually streaming information at the bottom of the screen, titles, headlines, and globe (graphic) information situated just above the crawler. To examine the effect of the presence of these potential distractors, Josephson and Holmes compared scanpaths over news stories displayed on a standard layout (no enhancements), a standard layout with the crawler, and a standard layout with crawler and headline enhancements. As expected, the presence of the crawler drew more visual attention to that portion of the screen, but did not diminish recall of the main story content. However, participants exposed to headline summaries were less likely to recall other story points, suggesting an information interference effect (e.g., when visual and aural information differed).

Similar studies can be performed to evaluate the effect of advertisements, either over typical commercial spots or during live broadcasts of sporting events, for example. American broadcasts of baseball and soccer incorporate advertisements into the video stream by either virtually compositing ads on green screen elements behind the batter's box, or on the scoreline atop the screen during football matches (e.g., during the FIFA Germany World Cup 2006 broadcasts). Eye tracking data add objective support to recall statistics of the effectiveness of such displays.

19.5 Web Pages

Web page layout design has received a good deal of interest lately, although eye tracking studies have not yet revealed groundbreaking guidelines for optimal design (whether they ever will is a valid question, one that most likely puts an undue expectation on eye movement analysis). However, several marketing-related research directions are worth mentioning.

Banner blindness, related perhaps to the phenomenon of inattentional blindness, received a great deal of early attention. It is difficult to suggest a single definitive reference to support, refute, or explain this phenomenon, although it appears to be generally accepted as a tendency of Web visitors to ignore banner ads, even when banners may contain actively sought information. For eye tracking evidence, the study conducted by the Nielsen Norman Group (NN/g) appears to have been given credence on the Internet (Nielson Norman Group, 2006; Nielson, 2006b).

Another of the Nielsen Norman Group's observations to be popularized describes search page scanning behavior. Similar to Eyetools'[1] "golden triangle" (Eyetools et al., 2006), NN/g reports an "F" pattern evident in aggregate scanning behavior and visible in Figure 19.4 (Nielson, 2006a). The implication of this scan pattern for marketing purposes is that viewers tend to look at the upper-left portion of search results (e.g., Google's) and ignore sponsors' links in the right column. Whether sponsors' links should be moved to the upper left is an open issue. If they are, their placement may disrupt search efficiency by impeding the typical visual search pattern to which users have now grown accustomed.

Whether current Web search results are presented in a manner evoking optimal visual search is an interesting research question. Rele and Duchowski (2005) used an eye tracker to study two different Web page search layouts. The first was a traditional list

---

[1] See: <http://www.eyetools.com/inpage/research_google_eyetracking_heatmap.htm>, last referenced July 7, 2006.

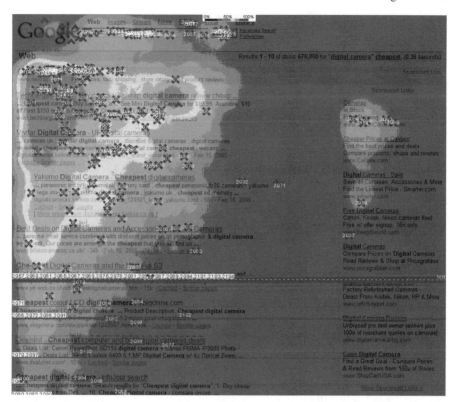

Fig. 19.4. Google's golden triangle (publicly available on the Internet, e.g., via Google's image search).

interface, where the list contained groups of elements delineated by categorical Web page information of title, summary, and URL. The second was a tabular rendition of the same information, where the distinct categorical elements were presented in separate columns instead of above and below each other. Figure 19.5 shows example aggregate "hotspots" from the study. Although there was no significant difference in performance measures of speed (search time for a given item) or accuracy (correct target identification), eye movements provided interesting insights.

Significant and sometimes task-dependent differences were found in the number of fixations over specific elements. Users performed one of two tasks: navigation or information gathering. Information gathering tasks were artificially created with the sought information present on one or more Web pages. In contrast, navigation tasks were designed without the sought information available, rather, it was meant to be inferred to exist on Web pages pointed to by the search results.

A significantly larger number of fixations were devoted to the summary category during navigation than when information gathering. This may suggest that difficult

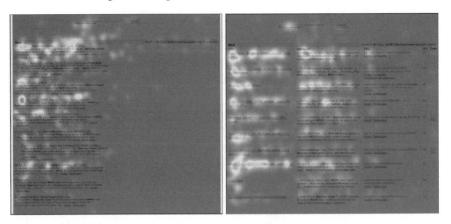

Fig. 19.5. Web search layout "hotspots": list (at left), tabular (at right). From Rele and Duchowski (2005) © 2005 Human Factors and Ergonomics Society. Reprinted by permission.

search tasks, in this case navigation that included searching for specific information, may require more careful reading of the link's summary before selection. A significantly larger number of fixations were also observed over the URL category with the list layout than tabular layout, this time independent of task. This finding would suggest that the URL is not as important to the viewer when it is possible to ignore, that is, displayed far enough away from the more informative title and summary information. However, the significance of this result may be partially tainted by fixation "spillover". It is difficult to reliably count deliberate fixations made on the URL element in the list layout. Due to the eye tracker's inherent limited accuracy, fixations intended over the summary element may have been mistakenly counted over the URL.

A cheap alternative to eye tracking visual attention, particularly over Web pages, may be the use of a mouse-contingent viewer such as the Enhanced Restricted Focus Viewer, or ERFV, developed by Tarasewich et al. (2005). The ERFV blurs regions outside a small window surrounding the mouse cursor in a similar manner to gaze-contingent displays (see Chapter 20). By doing so, the approach restricts visual attention to the cursor location. Recording mouse movements can then be used to visualize attention in a similar manner to displaying scanpaths. Tarasewich et al. suggest that such ERFV trajectories provide a good (statistically valid) estimate of the areas that subjects inspect when viewing Web pages.

One should consider the use of ERFV or mouse-contingent trajectories carefully, however, as other studies have raised considerable doubt concerning the method's validity (see Section 20.1.2). Although trajectories recorded by ERFV or similiar focus + context displays may appear similar to gaze scanpaths, they may alter performance metrics in a manner analogous to the Think-Aloud protocol. That is, task completion times with the ERFV may be slower, and perhaps more important, the ERFV may miss recording vital attentional behavior such as attention switching (Bednarik & Tukiainen, 2004, 2005). That is, the presence of the focus + context display may disrupt natural eye movements.

19.6 Product Label Design

Related to layout design are studies dealing with physical product labeling. Capturing eye movements over products arranged on real or mock supermarket shelves may require a head-mounted eye tracker (Li et al.'s (2006) *openEyes* project's low-cost head-mounted eye tracker may be very suitable for this purpose). Alternatively, label designs may be compared by viewing different designs on the desktop. Bojko et al. (2005) performed such a study where they tested product label designs, specifically drug labels. Eye tracking data helped explain the noted speed increase during

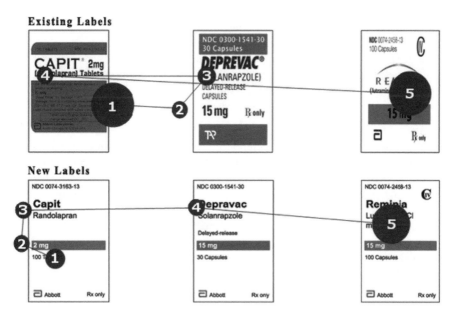

Fig. 19.6. Scanpaths over redesigned drug labels, with given task of finding the generic name *lutramine HCl monohydrate*. Notice the longer fixations on the existing designs compared to the new designs. From Bojko et al. (2005) © 2005 Human Factors and Ergonomics Society. Reprinted by permission.

visual search tasks over the newly redesigned labels: lowered search demands or lowered information-processing demands posed by the new design. For example, when searching for a label with the correct generic name (in this case *lutramine HCl monohydrate*), fixation duration was found to be significantly lower for the new designs than for existing designs (see Figure 19.6). This increased efficiency was attributed to the new labels' improved consistency in name formatting (font type and size). Eye movements helped corroborate increased visual search efficiency of the new label designs under other visual search tasks (e.g., searching for dosage formulation or strength), indicating the benefits of elemental changes such as better text positioning, clearer background, and more consistent font.

19.7 Summary and Further Reading

There are numerous opportunities for conducting eye tracking marketing studies. Copy testing, print advertising, and ad placement are suitable potential experiments that may be used to improve the impact of advertising materials. Unfortunately, evidence of eye tracking in market research is difficult to find. It may be that advertising companies do not wish to disclose the fact that eye trackers are being used. This may be perceived by the buying public as somehow being devious. Still, it is fairly safe to say that eye trackers are probably well known to marketing researchers and with improvements in technology will continue to be valuable tools in their work.

Possible sources of eye tracking research include scientific journals and professional conferences. One particular source that occasionally contains reports of eye tracking work is the *Journal of Advertising*. Another place to search for evidence of eye tracker use is the World Wide Web. A quick search of the Web reveals the following available reports.

- The Outdoor Advertising Association of America lists the following available research report: "The PRS Eye Tracking Studies: Validating Outdoor's Impact in the Marketplace" (see <http://www.oaaa.org/>, last accessed 05/31/02).
- The Institute of Behavioural Sciences (<http://ibs.derby.ac.uk/>) offers its eye tracking services for examining printed material (see <http://ibs.derby.ac.uk/research/eye.html>, last accessed 01/03/02).
- A conference announcement by the Advertising Research Foundation (ARF), Week Of Workshops (WOW), October 29–November 1, 2001, lists among its scheduled presentations the following: "New Ad Designs Capture Users' Eyes: A Case Study of Eye Tracking for CNET". (See <http://www.thearf.org/webpages/wow-2001/conference-tuesday-marketing.html>, last accessed 01/03/02.)

Thus it appears eye tracking market research is being conducted. The last example is particularly interesting because it shows a fairly new domain ripe for the study of advertising placement: the World Wide Web itself.

20

Computer Science

This chapter gives an overview of primarily interactive applications mainly developed by computer science researchers. Following the hierarchy of eye tracking systems given earlier, apart from diagnostic usability studies, this chapter focuses on two types of interactive applications: selective and gaze-contingent. The former approach uses an eye tracker as an input device, similar in some ways to a mouse. This type of ocular interaction is often studied by researchers involved in the fields of Human–Computer Interaction (HCI) and Computer-Supported Collaborative Work (CSCW). The latter gaze-contingent application is a type of display system wherein the information presented to the viewer is generally manipulated to match the processing capability of the human visual system, often matching foveo–peripheral perception in real-time. It should be noted that a good deal of previous work discussed in this chapter is based on conference proceedings.

20.1 Human–Computer Interaction and Collaborative Systems

Eye-based interactive systems have been presented at several SIGCHI conferences with a significant increase in the number of papers in recent years. Most of these papers have traditionally focused on interactive uses of eye trackers, although diagnostic applications have also begun to appear, particularly in the context of usability studies.

Interactive uses of eye trackers typically employ gaze as a pointing modality, e.g., using gaze in a similar manner to a mouse pointer. Prominent applications involve selection of interface items (menus, buttons, etc.), as well as selection of objects or areas in Virtual Reality (VR). A prototypical "real-world" application of gaze as an interactive modality is eye typing, particularly for handicapped users.

Other uses of gaze in the general field of human–computer interaction involve gaze as an indirect pointing modality, for example as a deictic reference in the context of collaborative systems, or as an indirect pointing aid in user interfaces. Diagnostic uses of eye trackers are becoming adopted for usability studies, i.e., testing the effectiveness of interfaces as evidenced by where users look on the display.

20.1.1 Classic Eye-Based Interaction

One of the first eye-based interactive systems, introduced by Jacob (1990), demonstrated an intelligent gaze-based informational display. In Jacob's system a text window would scroll to show information on visually selected items. The paper's title, "What You Look At Is What You Get," is a play on words on the common word processing/printing paradigm of What You See Is What You Get (WYSIWYG). Jacob's paper was an early paper describing the feasibility of a gaze-based interactive system in which the author discussed the possibility of using gaze in place of or in addition to a mouse pointer. The paper was one of the first to use video-based corneal reflection eye tracking technology interactively and is well known for its identification of an important problem in eye-based interactive systems: the Midas Touch problem. Essentially, if the eyes are used in a manner similar to a mouse, a difficulty arises in determining intended activation of foveated features (the eyes do not register button clicks!). That is, unlike a mouse with which a user signifies activation of an object by pressing a mouse button, with gaze pointing everything that a user looks at is potentially activated (and so in the Midas analogy unintentionally turns to gold). To avoid the Midas Touch problem, Jacob discusses several possible solutions including blinks, finally promoting the use of dwell time (of about 150–200 ms) to act as a selection mechanism.

At the same SIGCHI meeting in which Jacob's paper appeared, a graphical "self-disclosing" display was presented by Starker and Bolt (1990). This interactive system provided the user with gaze-controlled navigation in a three-dimensional graphics world. The graphical environment contained story world characters who responded in interesting ways to the user's gaze. Fixations activated object "behaviors", because the system would maintain and increase the user's visual interest level in an object. With increased interest, objects would blush and/or provide a verbal narrative. Unlike Jacob's use of dwell time, in this system dwell time was used to zoom into the graphics world.

At a more recent SIGCHI meeting, Tanriverdi and Jacob (2000) presented a new gaze-based interactive system, this time with gaze acting as a selective mechanism in VR. In this system Tanriverdi and Jacob compared eye-based interaction with hand-based interaction. The authors found that performance with gaze selection was significantly faster than hand pointing, especially in environments where objects were placed far away from the user's location. In contrast, no performance difference was found in "close" environments. Furthermore, although pointing speed may increase with gaze selection, there appears to be a cognitive tradeoff for this gain in efficiency: subjects had more difficulty recalling locations they interacted with when using gaze-based selection than when using hand selection.

20.1.2 Cognitive Modeling

In their paper presented at SIGCHI, Byrne et al. (1999) tested the arrangement of items during visual search of click-down menus. The authors contrasted two computational cognitive models designed to predict latency, accuracy, and ease of learning for a wide variety of HCI-related tasks: EPIC (Executive Process Interactive Control; Kieras and Meyer (1995)) and Adaptive Control of Thought-Rational (ACT-R; Anderson (2002)). The EPIC architecture provides a general framework for simulating a human interacting with her or his environment to accomplish a task (Hornof & Kieras, 1997). ACT-R is a framework for understanding human cognition whose basic claim is that cognitive skill is composed of production rules (Anderson, 1993). These models specifically predict, in different ways, the relationship between eye and mouse movement. With respect to click-down menus, the EPIC model predicts the following.

1. The distance covered by saccades is constant.
2. Eyes tend to overshoot the target with some regularity.
3. No mouse movement is initiated until the target is visually acquired.
4. Eye movement patterns are generally top-down and random.

In contrast, the ACT-R model predicts:

1. The distance covered by saccades is variable.
2. Eyes never overshoot the target.
3. Mouse movement generally follows saccades.
4. Eye movement patterns are exclusively top-down.

A visual search experiment was conducted where a target item was presented prior to display of a click-down menu. Based on the number of fixations, both models were supported because in some cases visual search exhibited exclusively top-down search (ACT-R) and in others both top-down and random patterns were observed (EPIC). Fixation positions, however, did not appear random, they did not appear to be strictly linear, and there appeared to be a preference for the first item in the menu. This particular study is informative for two reasons. First, it shows the importance of a good model (or lack of one) that can be used for designing user interfaces. If a model can be developed that successfully predicts some aspect of user activity, then this model may be used to design an interface that benefits the user by adapting to the user (instead of, for example, forcing the user to learn a new, possibly unintuitive style of interaction). The idea of modeling human behavior is an important concept in human–computer interaction. Second, the paper points out that eye tracking is an effective usability modality and that a new model for visual search over menus is needed, e.g., possibly a "noisy" version of top-down search.

Beyond keystroke-level modeling and their variants, eye tracking has recently been used in the field of psychology of programming: understanding the cognitive processes involved in debugging and comprehending computer programs. In an effort to develop an eye tracking methodology for studying program comprehension, Bednarik (2005) compared eye tracking to mouse-contingent display, i.e., blurring the display except in the neighborhood of the mouse cursor, as performed in the so-called Restricted Focus Viewer, or RFV technique.

The RFV technique has been suggested in the literature as an alternative to eye tracking, both because it is cheaper (it does not require any hardware in addition to the normal workstation), and because it is unobtrusive and accurate at the same time (in the sense that there is no measurement error (see Tarasewich et al. (2005) for a review of RFV and an ehanced method designed for Web page usability testing). However, it is reasonable to ask how much the blurring of most of the display affects normal viewing behavior and, thereby, possibly disrupting the observations made using the technique. Previous research has implied that the technique could be a valid tool, but the tasks where it has been used have been relatively simple. Bednarik used an eye tracker to repeat an experiment that had previously been performed with RFV and cast doubt on the latter's validity. Addressing the question in detail, controlled experiments showed that an eye tracker clearly outperformed RFV in terms of validity, given the particular task of program debugging. For other tasks, such as Web page usability testing, Tarasewich et al.'s Enhanced RFV technique may be useful, although probably not as a substitute for eye tracking (see Section 19.5).

Fig. 20.1. Scanning behavior during program debugging. From Uwano et al. (2006) © 2006 ACM, Inc. Reprinted by permission.

Studying the same cognitive processes (program comprehension), Uwano et al. (2006) provided interesting static visualiztions of dynamic eye movements (similar to those of Räihä et al. (2005)) during program debugging. Such visualizations can be used to depict differences between debugging strategies of expert and novice programmers and can thus lead to future models of behavior. An example is shown in Figure 20.1.

20.1.3 Universal Accessibility

Besides pointing in desktop and immersive (VR) displays, the archetypical gaze-based pointing application is eye typing. Eye typing is and has been a useful communication modality for the severely handicapped. Severely handicapped people, possessing an acute need for a communication system, may only be able to control their eyes. Most eye typing systems are implemented by presenting the user with a virtual keyboard, either on a typical computer monitor, or in some cases projected onto a wall. Based on analysis of tracked gaze, the system decides which letter the

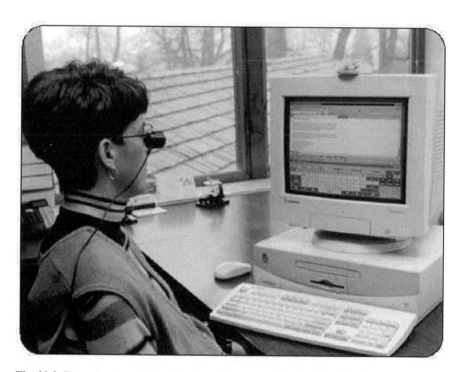

Fig. 20.2. Example of eye typing interface. Courtesy of Prentke Romich Company, Wooster, OH <http://www.prentrom.com/access/hm2000.html>. Reproduced with permission.

user is looking at and decides (e.g., by dwell time) whether to type this letter. The system may provide feedback to the user by either visual or auditory means, or a combination of both. Eye tracking systems may be either video-based or based on Electro-OculoGraphic (EOG) potential. An example eye typing interface is shown in Figure 20.2. Variations of traditional gaze-based pointing may be employed for subjects exhibiting difficulties fixating (e.g., locked-in syndrome). In such cases, if the eyes can only be moved in one direction, the eyes can be used as simple one- or two-way switches and the focus can be shifted from one item to another by using a method known as scanning. Using a combination of scanning and switching, the user can use one switch to change display focus by scanning across the display, then use another switch to indicate selection of the item currently in focus.

Fig. 20.3. *Sunny Day.* A drawing created solely with eye movements by an *EyeDraw* developer. Publicly available online at <http://www.cs.uoregon.edu/research/cm--hci/EyeDraw/> (last accessed 07/10/06).

Majaranta and Raiha (2002) provide an excellent survey of additional selection and feedback techniques employed in eye typing systems. Majaranta and Raiha have also been instrumental in the creation of COGAIN: Communication by Gaze Interaction,[1] a network of excellence supported by the European Commission's IST sixth framework program. COGAIN integrates cutting-edge expertise on interface technologies for the benefit of users with disabilities. The network aims to gather Europe's leading expertise in eye tracking integration with computers in a research project on assistive technologies for citizens with motor impairments. The COGAIN Web site provides a great deal of information, including (at the time of this writing) three downloadable

---

[1] See <http://www.cogain.org>.

eye communication systems (Dasher, Gazetalk, and UKO II) and various reports pertaining to the effort.

Although currently the United States lacks a far-reaching universal accessibility program such as COGAIN, some related successes have recently been reported. In particular, Hornof and Cavender (2005) developed *EyeDraw*, a gaze-based drawing program designed to enable children with severe motor impairments to draw with their eyes. The application is based on a simple drawing platform (e.g., draw lines, circles, etc.) but exposes the difficulty inherent with gaze-based control. As with the Midas Touch problem, the critical problem solved by the program is the interpretation of eye movements allowing users to click on buttons, choose drawing start and end points, and save and retrieve drawings. *EyeDraw* is freely available for download.[2] A screenshot with a drawing created solely with eye movements by an *EyeDraw* developer is shown in Figure 20.3.

20.1.4 Indirect Eye-Based Interaction

Gaze-based communication systems such as those featuring eye typing offer certain (sometimes obvious) advantages but are also problematic. Gaze may provide an often faster pointing modality than a mouse or other pointing device, especially if the targets are sufficiently large. However, gaze is not as accurate as a mouse because the fovea limits the accuracy of the measured point of regard to about 0.5 degree visual angle. Another significant problem is accuracy of the eye tracker. Following initial calibration, eye tracker accuracy may exhibit significant drift, where the measured point of regard gradually falls off from the actual point of gaze. Together with the Midas Touch problem, drift remains a significant problem for gaze input.

Zhai et al. (1999) take another approach to gaze-based interaction and test the use of gaze as a sort of predictive pointing aid rather than a direct effector of selection. This is a particularly interesting and significant departure from "eye pointing" because this strategy is based on the authors' assertion that loading of the visual perception channel with a motor control task seems fundamentally at odds with users' natural mental model in which the eye searches for, and takes in information, while coordinating with the hand for manipulation of external objects. In their paper on Manual Gaze Input Cascaded (MAGIC) Pointing, Zhai et al. present an acceleration technique where a (two-dimensional) cursor is warped to the vicinity of a fixated target. The acceleration is either immediate, tied to eye movement (liberal MAGIC pointing), or delayed, following the onset of mouse movement (conservative MAGIC pointing). The authors report that although the speed advantage is not obvious over manual (mouse) pointing (subjects tended to perform faster with the liberal method), almost all users subjectively felt faster with either new pointing technique.

Another interesting indirect gaze-based interaction was presented by Santella et al. (2006). Here, eye tracking is used to mark aesthetically important locations in an

---

[2] See <http://www.cs.uoregon.edu/research/cm-hci/EyeDraw/>.

image. Represented as a gaze-based content map, recorded fixations are then used to automatically crop the image. In a forced-choice ("which one looks better") user comparison, gaze-based croppings were preferred over originals as well as over automatically cropped images. However, professionally hand-cropped images were chosen over any of the other three methods.

20.1.5 Attentive User Interfaces (AUIs)

Enabling a computer application or other physical device awareness of gaze is a topic within the burgeoning area of attentive user interfaces. Vertegaal's (2003) special issue of the *Communications of the ACM* provided several articles falling under this heading. The issue reviews a number of devices that track a user's attention, including eye trackers, attentive phones, and attentive videoconferencing cameras. Developing the notion of AUIs further, Shell et al. (2004) introduced ECSGlasses and EyePliances: devices that used eye contact as an attentional cue to engage in a more sociable process of turn-taking with users. EyePliances relied on an eye detecting camera, the Eye Contact Sensor, for awareness of a user's gaze. Given this cue, the user could simply issue verbal instructions to a gaze-aware device. For example, one could turn on a lamp equipped with an Eye Contact Sensor by simply commanding "On!" Because the lamp could sense it was being looked at, it would "know" to turn itself on.

The requirement that EyePliances needed to be equipped with sensing cameras presented a technical limitation in terms of the camera's field of view. When multiple objects each equipped with an Eye Contact Sensor were placed within 80 degrees of visual angle from each other, an ambiguity arose in determining which object was being fixated. Rather than solving this problem, Smith et al. (2005) removed the camera from the EyePliance and fitted it on a lightweight head-mounted eyepiece. Their revised system for gaze-based *deixis*, or specification of a referent, e.g., "*that* object", by gaze, is termed ViewPointer, and is shown in Figure 20.4.

Any physical object can be augmented with a small infra-red (IR) tag. To detect eye contact, ViewPointer considers whether the reflection of the IR tag on the cornea appears central to the pupil. If so, the user is looking at the tag. Calibration is not needed because eye movements are not correlated to scene coordinates. In addition to simple eye contact sensing (i.e., a binary decision of whether a device is being looked at), Smith et al. also developed a novel encoding scheme that is used to identify each tag. Other data beyond tag identification can be transmitted visually to the ViewPointer camera at a rate of 14 bits per second (as of that version of the system; it is probable that this rate will increase in future revisions). Information can therefore be transmitted to the ViewPointer wearer as suggested by the usage scenario depicted in Figure 20.4. In this example, tags are mounted behind the poster in the store window. A URL tag can be transmitted along with tag id information, thereby providing the user with a pointer to relevant Web-based information that ViewPointer could transmit via Bluetooth to a handheld PDA or other browser-enabled device.

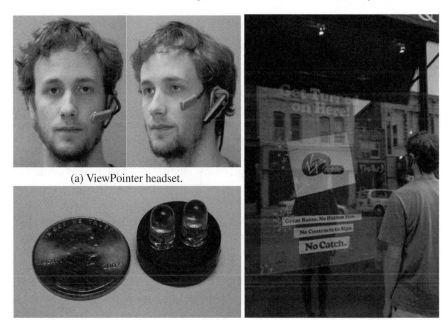

(a) ViewPointer headset.

(b) ViewPointer tag and US penny.   (c) ViewPointer usage.

Fig. 20.4. ViewPointer headset, tag, and usage: a user looks at a poster augmented with an invisible ViewPointer URL tag mounted behind the poster's logo. From Smith et al. (2005) © 2005 ACM, Inc. Reprinted by permission.

20.1.6 Usability

Besides the use of gaze for interactive means, diagnostic eye tracking is gaining acceptance within the HCI and usability communities (particularly practitioners) as another means to test usability of an interface. It is believed that eye movements can significantly enhance the observation of users' strategies while using computer interfaces (Goldberg & Kotval, 1999). Among various experiments, eye movements have been used to evaluate the grouping of tool icons, compare gaze-based and mouse interaction techniques (Sibert & Jacob, 2000), evaluate the organization of click-down menus, and to test the organization of Web pages.

In a usability study of Web pages, Goldberg et al. (2002) derive specific recommendations for a prototype Web interface tool. The authors discuss gaze-based evaluation of Web pages in which the system permits free navigation across multiple Web pages. This is a significant advancement for usability studies of Web browsers inasmuch as prior to this study recording of gaze over multiple Web pages has been difficult due to the synchronization of gaze over windows that scroll or hide from view. The authors describe their collection of the following dependent measures.

1. User actions such as key presses and mouse button clicks
2. Context-free eye movement measures such as fixations and dwell times

3. Context-sensitive eye movements such as dwell times within regions of interest

The key questions examined are whether eye movements are related to user actions, is navigation biased toward horizontal or vertical navigation, and whether there is any particular order to Web page scanning. Following task-level, screen-level, and object-level analyses, the authors report that users exhibit a preference for horizontal search across columns rather than searching within a column. Furthermore, recommendations are made for the left-hand placement of Web "portlets" requiring visibility, with the two most important portlets placed on top of the Web page.

A good deal many more gaze-based usability studies have been conducted. The above are but two examples, with the first being perhaps more relevant to user modeling than usability practice. However, since Karn et al.'s (1999) well-attended SGICHI 99 workshop, eye tracking appears to have gathered some momentum among usability practitioners. Following Jacob and Karn's (2003) review of eye movement metrics, Bojko (2005) presented useful eye tracking pointers garnered from practical experience. Bojko's study is mentioned in Chapter 14 where Guan et al.'s (2006) results on Retrospective Think-Aloud are also discussed. Results from Bojko et al.'s (2005) exemplary drug label experiment are covered in Chapter 19. A host of other practical issues were discussed in two more recent workshops aimed at practitioners, one again at SIGCHI 06 and another at UPA 06 (Renshaw et al., 2006; Webb & Renshaw, 2006). The recent findings of the "F" and "golden triangle" Web search scan patterns (attributed to Nielson Norman Group (2006) and Eyetools et al. (2006), respectively, see Chapter 19) suggest that eye tracking will continue to play an increasingly important role in usability investigations.

20.1.7 Collaborative Systems

Apart from interactive or diagnostic uses of eye movements, gaze can also be utilized to aid multiparty communication in collaborative systems. In the GAZE Groupware system, an LC Technologies eye tracker is used to convey gaze direction in a multiparty teleconferencing and document sharing system, providing a solution to two problems in multiparty mediated communication and collaboration: knowing who is talking to whom, and who is talking about what (Vertegaal, 1999). The system displays 2D images of remotely located participants in a VRML virtual world. These images rotate to depict gaze direction alleviating the problem of turn-taking in multiparty communication systems. Furthermore, a gaze-directed "lightspot" is shown over a shared document indicating the users' fixated regions and thereby providing a deictic ("look at this") reference. The system display is shown in Figure 20.5, with the the system interface shown in Figure 20.6. For further information, see <http://www.cs.queensu.ca/home/roel/gaze/home.html>.

Fig. 20.5. GAZE Groupware display. Courtesy of Roel Vertegaal.

(a) User interface.   (b) Eye tracking optics.

Fig. 20.6. GAZE Groupware interface. Courtesy of Roel Vertegaal.

20.2 Gaze-Contingent Displays

In general, eye-based interactive applications can be thought of as selective, because gaze is used to select or point to some aspect of the display, whether it is two-dimensional (e.g., desktop), collaborative, or immersive (such as a virtual environment). Mixing both directly interactive and indirectly "passive" usage styles of gaze are gaze-contingent displays. Here, gaze is used not so much as a pointing device, but rather as a passive indicator of gaze. Given the user's point of regard, a system can tailor the display so that the most informative details of the display are generated at the point of gaze, and degraded in some way in the periphery. The purpose of these displays is usually to minimize bandwidth requirements, as in video telephony applications, or in graphical applications where complex data sets cannot be fully displayed in real-time. Two main types of gaze-contingent applications are discussed: screen-based and model-based. The former deals with image (pixel) manipulation, and the latter is concerned with the manipulation of graphical objects or models. Both systems are generally investigated by researchers studying Computer Graphics (CG) and Virtual Reality (VR).

Human visual perception of digital imagery is an important contributing factor to the design of perceptually based image and video display systems. Human observers have been used in various facets of digital display design, ranging from estimation of corrective display functions (e.g., gamma function) dependent on models of human color and luminance perception, color spaces (e.g., CIE Lab color space), and image and video codecs. JPEG and MPEG both use quantization tables based on the notion of Just Perceptible Differences to quantize colors of perceptually similar hue (Wallace, 1991).

The idea of gaze-contingent displays is not new and dates back to early military applications (Kocian, 1987; Longridge et al., 1989). In the Super Cockpit Visual World Subsystem, Kocian considered visual factors including contrast, resolution, and color in the design of a head-tracked display. In their Simulator Complexity Testbed (SCTB), Longridge et al. included an eye-slaved ROI as a major component of the Helmet-Mounted Fiber Optic Display (HMFOD). This ROI provided a high-resolution inset in a low-resolution (presumably homogeneous) field that followed the user's gaze. The precise method of peripheral degradation was not described apart from the criteria of low resolution. However, the authors did point out that a smooth transition between the ROI and background was necessary in order to circumvent the possibility of a perceptually disruptive edge artifact.

Various gaze-contingent approaches have been proposed for foveal Region Of Interest (ROI)-based image and video coding (Stelmach & Tam, 1994; Nguyen et al., 1994; Kortum & Geisler, 1996; Tsumura et al., 1996). Often, however, these studies are based on automatically located image regions, which may or may not correspond to foveally viewed segments of the scene. That is, these studies do not necessarily employ an eye tracker to verify the ROI-based coding schemes. Instead, a figure-ground assumption is often used to argue for more or less obvious foveal candidates in the scene. This is a research area where either diagnostic eye movement studies can be used to corroborate the figure-ground assumption, or gaze may be used directly to display high-resolution ROIs at the point of regard in real-time (as in eye-based teleconferencing systems).

Instead of assuming a feature-based approach to foveal (high-resolution) encoding, an eye tracker can be used to directly establish the foveal ROI and a suitable image degradation scheme may be employed to render detail at the point of regard. This motivated research into finding a suitable image degradation scheme that would match foveal acuity.

20.2.1 Screen-Based Displays

When evaluating Gaze-Contingent Displays (GCDs), it is often necessary to distinguish between two main types of causalities: those affecting perception and those affecting performance. As a general rule, perception is more sensitive than performance. That is, it may be possible to degrade a display to a quite noticeable effect without necessarily degrading performance. In either case, one of the main difficulties that must be addressed is the latency of the system. Without predictive capabilities, most gaze-contingent displays will lag behind the user's gaze somewhat, usually by a constant amount of time proportional to both measurement of gaze (which may take up to the time it takes to display a single frame of video, typically 16 ms for a 60 Hz video-based tracker), and the subsequent time it takes to refresh the gaze-contingent display (this may take another 33 ms for a system with update rate of 30 frames per second).

Loschky and McConkie (2000) conducted an experiment on a gaze-contingent display investigating spatial, resolutional, and temporal parameters affecting perception and performance. Two key issues addressed by the authors are the timing of GCDs and the detectability of the peripherally degraded component of the GCD. That is, how soon after the end of an eye movement does the window need to be updated in order to avoid disrupting processing, and is there a difference between the window sizes and peripheral degradation levels that are visually detectable and those that produce behavioral effects? In all experiments, monochromatic photographic scenes were used as stimuli with a circular, high-resolution window surrounded by a degraded peripheral region. An example of Loschky and McConkie's GCD is shown in Figure 20.7(a). In one facet of the experiment, it was found that for an image change to go undetected, it must be started within 5 ms after the end of the eye movement. Detection likelihood rose quickly beyond that point. In another facet of the study concerning detection of peripheral degradation, results showed that the least peripheral degradation (inclusion of four of four possible levels) went undetected even at the smallest window size (2°), where the opposite was true with the highest level of degradation: it was quite detectable at even the largest window size (5°). The GCD was also evaluated in terms of performance effects, in the context of visual search and scene recall tasks. In the end it was found that the generation of an imperceptible GCD was quite difficult in comparison to the generation of a GCD that does not deteriorate performance. Although greater delays (e.g., 15 ms) and greater degradation (inclusion of only three of four possible levels) produce detectable visual artifacts, they appear to have minimal impact on performance of visual tasks when there is a 4.1° high-resolution area centered at the point of gaze.

Parkhurst et al. (2000) investigated behavioral effects of a two-region gaze-contingent display. A central high-resolution region, varying from 1 to 15 degrees, was presented at the instantaneous center of gaze during a visual search task. An example of Parkhurst et al.'s display is shown in Figure 20.7(b). Measures of reaction time, accuracy, and fixation durations were obtained during a visual search task. The authors' primary finding is that reaction time and accuracy co-vary as a function of the central region size. The authors note this as a clear indicator of a strategic speed/accuracy tradeoff where participants favor speed in some conditions and accuracy in others. For small central region sizes, slow reaction times are accompanied by high accuracy. Conversely, for large central region sizes, fast reaction times are accompanied by low accuracy. A secondary finding indicated that fixation duration varies

(a) From Loschky and McConkie (2000) © 2000 ACM, Inc. Reprinted by permission.

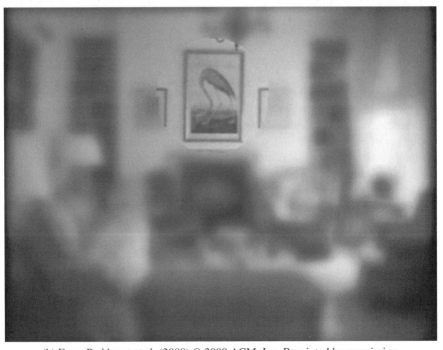

(b) From Parkhurst et al. (2000) © 2000 ACM, Inc. Reprinted by permission.

Fig. 20.7. Example gaze-contingent displays.

as a function of central region size. For small central region sizes, participants tend to spend more time examining each fixation than under normal viewing conditions. For large central regions, fixation durations tend to be closer to normal. In agreement with reaction time and accuracy, fixation duration is approximately normal (comparable to that seen for uniform resolution displays) with a central region size of 5°.

For screen-based VR rendering the work of Watson et al. (1997) is particularly relevant. The authors studied the effects of Level Of Detail (LOD) peripheral degradation on visual search performance. Both spatial and chrominance detail degradation effects were evaluated in Head-Mounted Displays (HMDs). To sustain acceptable frame rates, two polygons were texture mapped in real-time to generate a high-resolution inset within a low-resolution display field. The authors suggested that visual spatial and chrominance complexity can be reduced by almost half without degrading performance. More recently, Watson et al. (2004) used the same head-mounted display (but not head-tracked this time) to gain insights into peripheral LOD control beyond the perceptual threshold.

In an approach similar to Watson et al.'s, Reddy (1998) used a view-dependent LOD technique to evaluate both perceptual effects and system performance gains. The author reported a perceptually modulated LOD system which affords a factor of 4.5 improvement in frame rate.

Most of the above approaches result in a bi-resolution GCD, with the demarcation between the foveal disk region and peripheral resolution often purposefully made visible (thus without any inter-LOD blurring or averaging). Geisler and Perry (2002) proposed a method to generate completely arbitrary variable resolution displays. Their display depends on pyramidal preprocessing of the images prior to display (Geisler & Perry, 1998) (see Burt and Adelson (1983) for a detailed description of multiresolution pyramids with spatial filtering). Geisler and Perry's *Space Variant Imaging* software produces smooth nearly artifact-free images at high frame rates, but is limited to manipulation of spatial resolution. The software implementing this method on Windows platforms is freely available on-line.[3] Geisler and Perry's work is particularly significant for its separation of resolution degradation from image source. In image compositing parlance, this *switchmatte* operation makes gaze-contingent rendering immediately obvious. Simply preserve high-resolution pixels only at matte locations with $\alpha = 1$ and map pixels at lower matte luminance levels to lower-resolution pixels (e.g., from a bank of preprocessed images).

Because pyramidal reconstruction schemes draw pixel data from multiple levels of resolution, pyramidal image synthesis provides a smoothly degraded, convincing visualization of the human visual system, termed *foveation*. A particularly popular pyrmadical approach relies on image decomposition via the Discrete Wavelet Transform (DWT) with selective coefficient scaling and decimation prior to reconstruction

---

[3] <http://fi.cvis.psy.utexas.edu>.

(Chang et al., 2000; Duchowski, 2000). Using the DWT for smooth image resolution degradation, images demonstrating three acuity mapping functions are shown in Figure 20.8. For demonstration purposes, the *CNN* image was processed with an artificially placed ROI over the anchor's right eye and another over the "timebox" found in the bottom-right corner of the image. Haar wavelets were used to accentuate the visibility of resolution bands (see Section 4.2). Figure 20.8b, d, and f show the extent of wavelet coefficient scaling in frequency space. The middle row shows a reconstructed image where resolution drops off smoothly, matching visual acuity for a particular screen display at a particular viewing distance. Provided appropriate wavelet filters can be found, reconstruction exactly matches linear mipmapping.

Recently, Cöltekin (2006) used a pyramid-based LOD management method for close-range stereo photogrammetric rendering. Meanwhile, Böhme et al.'s (2006) GCD goes beyond spatial resolution degradation by degrading temporal content as well as spatial content. Spatiotemporal degradation extends Geisler and Perry's pyramidal preprocessing approach into the temporal dimension.

Although recent GCD implementations have yielded new insights into perception, most of these approaches still rely on somewhat restrictive computational approaches, i.e., either software-based pyramidal image reconstruction (as in Böhme et al.'s spatiotemporal video degradation) or a texture-mapped rendering with a limited number of textures (as in Watson et al.'s dual viewport composition). The limitation of these techniques surfaces either in their limited speed or display characteristics. Böhme et al. claim 30 frames per second performance but have previously reported an average system latency of 60 ms (Dorr et al., 2005). Watson et al.'s dual viewports limited the display to two distinct regions. It is unlikely that this approach lends itself to the simulation of arbitrary visual fields. Indeed, Reddy (2001) noted that practically all perceptually based work (up to that point) had used a small set of presimplified versions of an object from which to choose to render in a view-dependent manner. It appears that this is still the predominant approach without the apparent employment of the GPU for image synthesis. Reddy's *Percept* visualization performs a per-pixel calculation of the pixel's spatial frequency based on angular velocity and eccentricity whereas Cöltekin's *Foveaglyph* builds a pyramid of scaled images.

Due to recent advancements in computer hardware, gaze-contingent imaging research has appeared where image processing operations are performed in real-time, either by dedicated image processing hardware, or by more general-purpose graphics engines. In a recent example of dedicated hardware-accelerated eye-movement controlled image coding, Bergström (2003) used a DCT-based image codec to achieve real-time image compression and display.

(a) Haar linear mapping.

(b) Linear mapping.

(c) Haar HVS mapping.

(d) HVS mapping.

(e) Haar nonlinear mapping.

(f) Nonlinear mapping.

Fig. 20.8. Image reconstruction and wavelet coefficient resolution mapping (assuming 50 dpi screen resolution). Reprinted from Duchowski (2000) with permission © 2000 IEEE.

A GPU-based gaze-contingent display is now available[4] with minimal progamming effort. A short GLSL program allows simulation of arbitrary visual fields, as inspired by Geisler and Perry (2002), but degraded chromatically as well as spatially. The approach is a natural extension of Nikolov et al.'s (2004) and Duchowski's (2004) independent introduction of multitexturing approaches.

20.2.2 Model-Based Graphical Displays

As an alternative to the screen-based peripheral degradation approach, model-based methods aim at reducing resolution by directly manipulating the model geometry prior to rendering. The technique of simplifying the resolution of geometric objects as they recede from the viewer, as originally proposed by Clarke (1976), is now standard practice, particularly in real-time applications such as VR (Vince, 1995). Clarke's original criterion of using the projected area covered by the object for descending the object's LOD hierarchy is still widely used today. However, as Clarke suggested, the LOD management typically employed by these polygonal simplification schemes relies on precomputed fine-to-coarse hierarchies of an object. This leads to uniform, or *isotropic*, object resolution degradation.

Ohshima et al. (1996) proposed a gaze-contingent model-based adaptive rendering scheme where three visual characteristics were considered: central/peripheral vision, kinetic vision, and fusional vision. The LOD algorithm generated isotropically degraded objects at different visual angles. Although the use of a binocular eye tracker was proposed, the system as discussed used only head tracking as a substitute for gaze tracking.

Isotropic object degradation is not always desirable, especially when viewing large objects at close distances. In this case, traditional LOD schemes will display an LOD mesh at its full resolution even though the mesh may cover the entire field of view. Because acute resolvability of human vision is limited to the foveal 5°, object resolution need not be uniform. This is the central tenet of gaze-contingent systems.

Numerous multiresolution mesh modeling techniques suitable for gaze-contingent viewing have recently been developed (Zorin & Schröder, 2000). Techniques range from multiresolution representation of arbitrary meshes to the management of LOD through peripheral degradation within an HMD where gaze position is assumed to coincide with head direction (Lindstrom et al., 1996; MacCracken & Joy, 1996; Hoppe, 1997; Zorin et al., 1997; Schmalstieg & Schaufler, 1997). Although some of these authors address view and gaze dependent object representation, few results concerning display speedup are as yet available showing successful adaptation of these techniques within a true gaze-contingent system, i.e., one where an eye tracker is employed. Due to the advancements of multiresolution modeling techniques and to the increased affordability of eye trackers, it is now becoming feasible to extend the

---

[4] <http://andrewd.ces.clemson.edu/gcd/>.

LOD approach to gaze-contingent displays, where models are rendered nonisotropically.

An early example of a nonisotropic model-based gaze-contingent system, where gaze direction is directly applied to the rendering algorithm, was presented by Levoy and Whitaker (1990). The authors' spatially adaptive near-real-time ray tracer for volume data displayed an eye-slaved ROI by modulating both the number of rays cast per unit area on the image plane and the number of samples drawn per unit length along each ray as a function of local retinal acuity. The ray-traced image was sampled by a nonisotropic convolution filter to generate a 12° foveal ROI within a 20° mid-resolution transitional region. Based on preliminary estimates, the authors suggested a reduction in image generation time by a factor of up to 5. An NAC Eye Mark eye tracker was used to determine the user's POR while viewing a conventional 19 in. TV monitor. A chin rest and immobilization strap were used to eliminate the need for head tracking.

For environments containing significant topological detail, such as virtual terrains, rendering with multiple levels of detail, where the level is based on user position and gaze direction, is essential to provide an acceptable combination of surface detail and frame rate (Danforth et al., 2000). Recent work in this area has been extensive. Particularly impressive is Hoppe's (1998) view-dependent progressive mesh framework, where spatial continuity is maintained through structure design, and temporal continuity is maintained by geomorphs.

Danforth et al. (2000) used an eye tracker as an indicator of gaze in a gaze-contingent multiresolution terrain navigation environment. A surface, represented as a quadrilateral mesh, was divided into fixed-size (number of vertices) subblocks, allowing rendering for variable LOD on a per-subblock basis. From a fully detailed surface, lower levels of resolution were constructed by removing half of the vertices in each direction and assigning new vertex values. The new values were averages of the higher resolution values. Resolution level was chosen per sub-block, based on viewer distance. The resolution level was not discrete; it was interpolated between the precomputed discrete levels to avoid "popping" effects. The terrain, prior to gaze-contingent alteration, is shown in Figure 20.9. Rocks in the terrain are rendered by billboarding; i.e., images of rocks from the Pathfinder mission to Mars were rendered onto 2D transparent planes that rotate to maintain an orientation orthogonal to the viewer. Two views of the gaze-contingent environment (shown rendered and in wireframe) are seen in Figures 20.10 and 20.11. To exaggerate the gaze-contingent effect, in this environment, fractal mountains appear and disappear from view, based on direction of gaze. Notice also, in Figures 20.10a and b, the increased resolution (number of quads) below the gaze vector. The images in the figure are snapshots of the scene images generated by the eye tracker; i.e., what is seen by the operator (the point of regard crosshair, coordinates, and video frame timecode) is not seen by the viewer immersed in the environment.

(a) Wireframe.

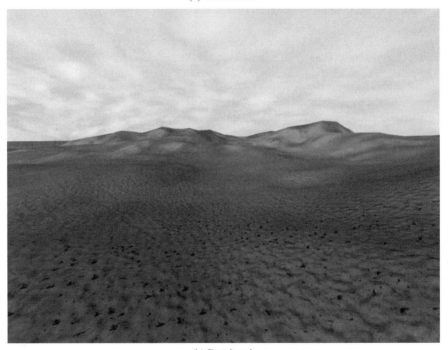

(b) Rendered.

Fig. 20.9. Fractal terrain for gaze-contingent virtual environment. Courtesy of Bob Danforth.

(a) Looking left.

(b) Looking right.

Fig. 20.10. Fractal terrain: gaze-contingent rendering (wireframe).

(a) Looking left.

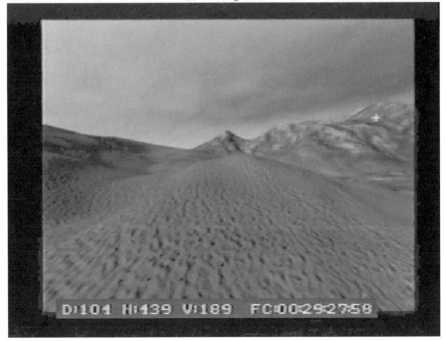

(b) Looking right.

Fig. 20.11. Fractal terrain: gaze-contingent rendering.

More recent work on gaze-contingent LOD modeling has been carried out by Luebke and Erikson (1997). The authors present a view-dependent LOD technique suitable for gaze-contingent rendering. Although simplification of individual geometric
objects is discussed in their work, it appears the strategy is ultimately directed toward solving the interactive "walkthrough" problem (Funkhouser & Séquin, 1993).
In this application, the view-dependent LOD technique seems more suitable to the
(possibly) gaze-contingent rendering of an entire scene or environment. Recently,
the authors have developed a gaze-directed LOD technique to facilitate the gaze-
contingent display of geometric objects (Luebke et al., 2000). To test their rendering
approach the authors employed a table-mounted monocular eye tracker to measure
the viewer's real-time location of gaze over a desktop display. This work shows the
feasibility of employing an eye tracker, however the implementation framework used
by the authors lacked a head tracker and required a chin rest to ensure tracker accuracy.

A new object-based LOD method has been developed by Murphy and Duchowski
(2001). The technique is similar to Luebke and Erikson's and to Ohshima et al.'s,
where objects are modeled for gaze-contingent viewing. Unlike the approach of
Ohshima et al., resolution degradation is applied nonisotropically; i.e., objects are
not necessarily degraded uniformly. The spatial degradation function for LOD selection differs significantly from the area-based criterion originally proposed by Clarke.
Instead of evaluating the screen coverage of the projected object, the degradation
function is based on the evaluation of visual angle in world coordinates. System performance measurements are reported, obtained from experiments using a binocular
eye tracker built into an HMD. Tracking software obtains helmet position and orientation in real-time and calculates the direction of the user's gaze. The geometric
modeling technique developed for the purpose of gaze-contingent rendering includes
an integrated approach to tiling, mapping, and remeshing of closed surfaces. A three-
dimensional spatial degradation function was obtained from human subject experiments in an attempt to imperceptibly display spatially degraded geometric objects.
System performance measurements indicate an approximate overall tenfold average
frame rate improvement during gaze-contingent viewing. Two frames during gaze-
contingent viewing of one geometric model are shown in Figure 20.12.

Another interesting approach to gaze-contingent modeling for real-time graphics
rendering was taken by O'Sullivan and Dingliana (2001) and O'Sullivan et al.
(2002). Instead of degrading the resolution of peripherally located geometric objects, O'Sullivan and Dingliana (2001) considered a degradable collision handling
mechanism to effectively limit object collision resolution outside a foveal Region Of
Interest (ROI). When the viewer is looking directly at a collision, it is given higher
(computational) priority than collisions occurring in the periphery. Object collisions
given higher priority are allocated more processing time so that the contact model
and resulting response are more believable. The calculated collision response accuracy (i.e., the precision of the calculated collision point between two rigid objects) is
directly dependent on the level of collision detection detail. A lower level of detail

(a) Original image.   (b) Looking at head.   (c) Looking at feet.

Fig. 20.12. Gaze-contingent viewing of *Isis* model. Courtesy of Hunter Murphy.

results in less physically accurate physics and objects not touching when they bounce, leaving a potentially perceivable gap.

O'Sullivan and Dingliana (2001) describe psychophysical experiments to measure the perceptibility of coarsely simulated collisions in the periphery. The authors tested effects of factors such as eccentricity, separation, presence and number of similar and dissimilar distractors, causality, and physics on participants' perception of collisions. To test viewers' sensitivity to collision resolution (size of gap between colliding objects) at eccentricity, O'Sullivan and Dingliana tested whether the ability of viewers to detect anomalous collisions, in this case colliding objects that do not touch each other but leave a gap, decreases with increasing eccentricity of the collision point. Collisions were presented at five eccentricities of $1.4°$, $2.9°$, $4.3°$, $5.7°$, and $7.2°$ visual angle. The authors note a significant fall-off in detection accuracy with eccentricity (at about $4°$ visual angle).

Based on their previous psychophysical findings, O'Sullivan et al. (2002) developed a gaze-contingent collision handling system. Two variants of the gaze-contingent system were compared, each containing a high-priority ROI wherein collisions were processed at higher resolution than outside the ROI. In the tracked case, the high-priority ROI was synchronized to the viewer's tracked gaze position with an SMI

EyeLink eye tracker, and in the random case, the ROI position was determined randomly every five frames. An example of the graphics simulation with highlighted ROI and subject wearing an eye tracker is shown in Figure 20.13. O'Sullivan et al. report an overall improvement in the perception of the tracked simulation.

(a) Important collisions (e.g., those close to the viewer's fixation position) are processed first.

(b) An eye-tracker is used to determine the viewer's point of fixation.

Fig. 20.13. Gaze-contingent collision modeling. Images courtesy of C. O'Sullivan and J.Dingliana.

20.3 Summary and Further Reading

Several classes of eye tracking applications were presented falling generally within the domain of computer wcience and human–computer interaction. HCI-related studies have included some of the first well-known adaptations of eye trackers to computer-based systems. This trend will probably continue for some time. Interactive eye tracking systems including those featured in Computer-Supported Collaborative Work (CSCW) will most likely continue to be explored for directly interactive support (e.g., gaze pointing) as well as for indirect interaction (e.g., gaze-assisted pointing). However, the recent interest in diagnostic uses of eye trackers, especially in usability studies, is also expected to flourish.

For an excellent review of Gaze-Contingent Multi-Resolution Displays, or GCM-RDs, see Reingold et al. (2002) as well as Parkhurst and Niebur (2002). The remainder of gaze-contingent work described in this chapter mostly emanates from research in computer graphics. Gaze-contingent displays described here are usually interactive real-time examples of adoption of eye trackers into graphical displays. See O'Sullivan et al. (2002) for a review of eye tracking work in interactive graphics. Here too, eye trackers are becoming noticed for their diagnostic contributions. The recent ACM SIGGRAPH Campfire on Perceptually Adaptive Graphics (McNamara & O'Sullivan, 2001) suggests that general perceptual issues are becoming particularly important in computer graphics research. Traditional graphics algorithms (e.g., ray-tracing, radiosity) are beginning to include perceptually based improvements to further speed up processing time. Eye trackers will certainly be able to contribute in a diagnostic capacity to improve the fidelity of perceptually enhanced graphical imagery.

For further reading, the two best sources of current research in HCI and computer graphics are the annual ACM SIGCHI and ACM SIGGRAPH conferences and the related ACM journal articles in the *Transactions on Computer–Human Interaction* and the *Transactions on Graphics*. Work specifically related to virtual reality can be found in the ACM VRST and IEEE VR conferences as well as in the journal *Presence*.

Work related to perception can be found in the recently formed *Transactions on Applied Perception (TAP)*, *Transactions on Multimedia Computing, Communications, and Applications (TOMCCAP)*, and the proceedings of the Symposium on Applied Perception in Graphics and Visualization (APGV).

Conclusion

Eye tracking has sometimes been referred to as a technology in search of an application. Through the presentation of examples in this text, which is by no means exhaustive, it appears that many interesting applications have now been found. There is a good deal of opportunity for meaningful research.

It has also been said that eye trackers are difficult to set up, use, and that they are unwieldy and expensive. There may be some truth in this observation. Through ongoing improvements to the technology, eye trackers are becoming much easier to use. Video-based eye trackers in particular are becoming relatively inexpensive, quite accurate, and fairly easy to use. The current "Holy Grail" of eye tracking research is a calibration-free eye tracker. Several research centers are pursuing this goal. With the use of multiple video cameras and computer vision techniques, this goal may be achieved very soon.

Setup and design of supporting computer programs and methodologies for eye tracking studies may still require some expertise and thus may pose some challenges. It is hoped that this book has in some way been useful in addressing these challenges. The contents of the text primarily deal with the computational "back-end" methodological issues, e.g., program design for graphical and virtual reality displays and subsequent eye movement collection and analysis. Often a complete eye tracking study may require collaboration with members from several traditionally distinct disciplines. What has seemed to work in the classroom may also work in the field, the lab, or on campus, and that is the formation of interdisciplinary teams. The assembly of members from say psychology, marketing, industrial engineering, and computer science may be the most effective strategy for conducting effective eye tracking research.

References

Allopenna, P. D., Magnuson, J. S., & Tanenhaus, M. K. (1998). Tracking the Time Course of Spoken Word Recognition Using Eye Movements: Evidence for Continuous Mapping Models. *Journal of Memory and Language, 38*(4), 419–439.

Ancman, C. E. G. (1991). Peripherally Located CRTs: Color Perception Limitations. In *National Aerospace and Electronics Conference (NAECON)* (pp. 960–965). Dayton, OH.

Anders, G. (2001). Pilot's Attention Allocation During Approach and Landing Eye- and Head-Tracking Research in an A330 Full Flight Simulator. In *International Symposium on Aviation Psychology (ISAP)*. Columbus, OH. (URL: <http://www.geerdanders.de/literatur/2001_ohio.html>. Last accessed December 2001.)

Anderson, J. R. (1993). *Rules of the Mind.* Hillsdale, NJ: Lawrence Erlbaum.

Anderson, J. R. (2002). *Personal Communique.*

Anliker, J. (1976). Eye Movements: On-Line Measurement, Analysis, and Control. In R. A. Monty & J. W. Senders (Eds.), *Eye Movements and Psychological Processes* (pp. 185–202). Hillsdale, NJ: Lawrence Erlbaum.

Arbib, M. A. (Ed.). (1995). *The Handbook of Brain Theory and Neural Networks.* Cambridge, MA: MIT Press.

Asaad, W. F., Rainer, G., & Miller, E. K. (2000). Task-Specific Neural Activity in the Primate Prefrontal Cortex. *Neurophysiology, 84*, 451–459.

Ashmore, M., Duchowski, A. T., & Shoemaker, G. (2005). Efficient Eye Pointing with a Fisheye Lens. In *Proceedings of Graphics Interface.* Victoria, BC, Canada.

Bahill, A. T., Clark, M., & Stark, L. (1975). The Main Sequence, a Tool for Studying Human Eye Movements. *Mathematical Biosciences, 24*(3/4), 191–204.

Balk, S. A., Moore, K. S., Steele, J. E., Spearman, W. J., & Duchowski, A. T. (2006). Mobile Phone Use in a Driving Simulation Task: Differences in Eye Movements. In *Vision Sciences Society (Posters)*. Sarasota, FL. (URL: <http://journalofvision.org/6/6/872/>. Last accessed July 2006.)

Ballard, D. H., Hayhoe, M. M., & Pelz, J. B. (1995). Memory Representations in Natural Tasks. *Journal of Cognitive Neuroscience*, *7*(1), 66–80.

Bass, M. (Ed.). (1995). *Handbook of Optics: Fundamentals, Techniques, & Design* (2nd ed.). New York: McGraw-Hill, sponsored by the Optical Society of America.

Becker, W. (1989). Metrics. In R. H. Wurtz & M. E. Goldberg (Eds.), *The Neurobiology of Saccadic Eye Movements* (pp. 13–68). New York: Elsevier Science BV (Biomedical Division).

Bednarik, R. (2005). *Towards an Eye-Tracking Methodology for Studies of Programming*. Unpublished master's thesis, University of Joensuu, Joensuu, Finland.

Bednarik, R., & Tukiainen, M. (2004). Visual Attention Tracking During Program Debugging. In *NordiCHI '04: Proceedings of the Third Nordic Conference on Human-Computer Interaction* (pp. 331–334). Tampere, Finland.

Bednarik, R., & Tukiainen, M. (2005). Effects of Display Blurring on the Behavior of Novices and Experts During Program Debugging. In *CHI '05: Extended Abstracts on Human Factors in Computing Systems* (pp. 1204–1207). Portland, OR.

Bergström, P. (2003). *Eye-movement Controlled Image Coding*. Unpublished doctoral dissertation, Linköping University, Linköping, Sweden.

Bertera, J. H., & Rayner, K. (2000). Eye Movements and the Span of the Effective Stimulus in Visual Search. *Perception & Psychophysics*, *62*(3), 576–585.

Bloom, F. E., & Lazerson, A. (1988). *Brain, Mind, and Behavior* (2nd ed.). New York: W. H. Freeman and Company.

Boff, K. R., & Lincoln, J. E. (Eds.). (1988). *Engineering Data Compendium: Human Perception and Performance*. Wright-Patterson AFB, OH: USAF Harry G. Armstrong Aerospace Medical Research Laboratory (AAMRL).

Böhme, M., Dorr, M., Martinetz, T., & Barth, E. (2006). Gaze-Contingent Temporal Filtering of Video. In *Eye Tracking Research & Applications (ETRA) Symposium* (pp. 109–116). San Diego.

Bojko, A. (2005). Eye Tracking in User Experience Testing: How to Make the Most of It. In *Proceedings of UPA '05*. Montréal, Canada.

Bojko, A., Gaddy, C., Lew, G., Quinn, A., & Israelski, E. (2005). Evaluation of Drug Label Designs Using Eye Tracking. In *Proceedings of the Human Factors and Ergonomics Society, 49th Annual Meeting*. Orlando, FL.

Brinkmann, R. (1999). *The Art and Science of Digital Compositing*. New York: Morgan Kaufmann.

Broadbent, D. E. (1958). *Perception and Communication*. Oxford: Pergamon Press.

Burt, P. J., & Adelson, E. H. (1983). The Laplacian Pyramid as a Compact Image Code. *IEEE Transactions on Communications*, *31*(4), 532–540.

Buswell, G. T. (1935). *How People Look At Pictures*. Chicago: University of Chicago Press.

Butenhof, D. R. (1997). *Programming with POSIX Threads*. Reading, MA: Addison-Wesley Longman.

Byrne, M. D., Anderson, J. R., Douglass, S., & Matessa, M. (1999). Eye Tracking the Visual Search of Click-Down Menus. In *Human Factors in Computing Systems: CHI '99 Conference Proceedings* (pp. 402–409). ACM Press.

Carmi, R., & Itti, L. (2006). Causal Saliency Effects During Natural Vision. In *Proceedings of Eye Tracking Research & Applications (ETRA)* (pp. 11–18). New York: ACM Press.

Carpendale, M. S. T., Cowperthwaite, D. J., & Fracchia, F. D. (1995). 3-dimensional pliable surfaces: For the effective presentation of visual information. In *UIST '95: Proceedings of the 8th annual ACM Symposium on User Interface and Software Technology* (pp. 217–226). ACM Press.

Carpenter, R. H. S. (1977). *Movements of the Eyes.* London: Pion.

Chang, E.-C., Mallat, S., & Yap, C. (2000). Wavelet Foveation. *Journal of Applied Computational Harmonic Analysis, 9*(3), 312–335.

Chapman, P. R., & Underwood, G. (1998). Visual Search of Dynamic Scenes: Event Types and the Role of Experience in Viewing Driving Situations. In G. Underwood (Ed.), *Eye Guidance in Reading and Scene Perception* (pp. 369–394). Oxford: Elsevier.

Clark, M. R., & Stark, L. (1975). Time Optimal Behavior of Human Saccadic Eye Movement. *IEEE Transactions on Automatic Control, 20,* 345–348.

Clarke, J. H. (1976). Hierarchical Geometric Models for Visible Surface Algorithms. *Communications of the ACM, 19*(10), 547–554.

Cöltekin, A. (2006). *Foveation for 3D Visualization and Stereo Imaging.* Unpublished doctoral dissertation, Helsinki University of Technology, Helsinki, Finland. (ISBN 951-22-8017-5, URL: <http://lib.tkk.fi/Diss/2006/isbn9512280175/>. Last accessed July 2006.)

Comer, D. E., & Stevens, D. L. (1993). *Internetworking With TCP/IP Vol III: Client-Server Programming And Applications* (BSD Socket Version ed.). Englewood Cliffs, NJ: Prentice-Hall.

Comer, D. E., & Stevens, D. L. (2001). *Internetworking With TCP/IP Vol III: Client-Server Programming And Applications* (Linux/POSIX Socket Version ed.). Upper Saddle River, NJ: Prentice-Hall.

Coolican, H. (1996). *Introduction to Research Methods and Statistics in Psychology* (2nd ed.). London: Hodder & Stoughton.

Cooper, R. M. (1974). The Control of Eye Fixation by the Meaning of Spoken Language: A New Methodology for the Real-Time Investigation of Speech Perception, Memory, and Language Processing. *Cognitive Psychology, 6,* 84–107.

Cornsweet, T. N. (1970). *Visual Perception.* New York: Academic Press.

Crane, H. D. (1994). The Purkinje Image Eyetracker, Image Stabilization, and Related Forms of Stimulus Manipulation. In D. H. Kelly (Ed.), *Visual Science and Engineering: Models and Applications* (pp. 13–89). New York: Marcel Dekker.

Crane, H. D., & Steele, C. M. (1985). Generation-V Dual-Purkinje-Image Eyetracker. *Applied Optics, 24,* 527–537.

Crichton, M. W. (1981). *Looker [Motion Picture].* Warner Brothers (USA).

Crundall, D. E., Underwood, G., & Chapman, P. R. (1998). How Much Do Novice Drivers See? The Effects of Demand on Visual Search Strategies in Novice and Experienced Drivers. In G. Underwood (Ed.), *Eye Guidance in Reading and Scene Perception* (pp. 395–418). Oxford: Elsevier.

Danforth, R., Duchowski, A., Geist, R., & McAliley, E. (2000). A Platform for Gaze-Contingent Virtual Environments. In *Smart Graphics (Papers from the 2000 AAAI Spring Symposium, Technical Report SS-00-04)* (pp. 66–70). Menlo Park, CA.

Davson, H. (1980). *Physiology of the Eye* (4th ed.). New York: Academic Press.

Deutsch, J. A., & Deutsch, D. (1963). Attention: Some Theoretical Considerations. *Psychological Review*, *70*(1), 80–90.

De Valois, R. L., & De Valois, K. K. (1988). *Spatial Vision.* New York: Oxford University Press.

Dishart, D. C., & Land, M. F. (1998). The Development of the Eye Movement Strategies of Learner Drivers. In G. Underwood (Ed.), *Eye Guidance in Reading and Scene Perception* (pp. 419–430). Oxford: Elsevier.

Dix, A., Finlay, J. E., Abowd, G. D., & Beale, R. (2004). *Human-Computer Interaction* (2nd ed.). Harlow, England: Pearson Education Limited.

Dodge, R. (1907). An Experimental Study of Visual Fixation. *Psychological Monograph*, *8*(4).

Doll, T. J. (1993). Preattentive Processing in Visual Search. In *Proceedings of the Human Factors and Ergonomics Society, 37th Annual Meeting* (pp. 1291–1249). Santa Monica, CA.

Doll, T. J., Whorter, S. W., & Schmieder, D. E. (1993). Simulation of Human Visual Search in Cluttered Backgrounds. In *Proceedings of the Human Factors and Ergonomics Society, 37th Annual Meeting* (pp. 1310–1314). Santa Monica, CA.

Dorr, M., Böhme, M., Martinetz, T., & Barth, E. (2005). Visibility of Temporal Blur on a Gaze-Contingent Display. In *Applied Perception, Graphics & Visualization (APGV) Symposium.* Coruña, Spain.

Dowling, J. E., & Boycott, B. B. (1966). Organization of the Primate Retina: Electron Microscopy. *Proceedings of the Royal Society (London)*, *166*, 80–111. (Series B)

Doyal, J. A. (1991). *Spatial Summation for Color in the Peripheral Retina.* Unpublished master's thesis, Wright State University.

Drury, C. G., Gramopadhye, A. K., & Sharit, J. (1997). Feedback Strategies for Visual Inspection in Airframe Structural Inspection. *International Journal of Industrial Ergonomics*, *19*, 333–344.

Duchowski, A., Medlin, E., Cournia, N., Gramopadhye, A., Melloy, B., & Nair, S. (2002). 3D Eye Movement Analysis for VR Visual Inspection Training. In *Eye Tracking Research & Applications (ETRA)* (pp. 103–110,155). New Orleans: ACM.

Duchowski, A., Medlin, E., Gramopadhye, A., Melloy, B., & Nair, S. (2001). Binocular Eye Tracking in VR for Visual Inspection Training. In *Virtual Reality Software & Technology (VRST)* (pp. 1–9). Banff, AB, Canada.

Duchowski, A. T. (2000). Acuity-Matching Resolution Degradation Through Wavelet Coefficient Scaling. *IEEE Transactions on Image Processing*, *9*(8), 1437–1440.

Duchowski, A. T. (2004). Hardware-Accelerated Real-time Simulation of Arbitrary Visual Fields. In *Eye Tracking Research & Applications (ETRA) Symposium* (p. 59). San Antonio, TX.

d'Ydewalle, G., Desmet, G., & Van Rensbergen, J. (1998). Film Perception: The Processing of Film Cuts. In G. Underwood (Ed.), *Eye Guidance in Reading and Scene Perception* (pp. 357–368). Oxford: Elsevier.

Eyetools, Enquiro, & Did-it. (2006). *Eyetools, Enquiro, and Did-it uncover Search's Golden Triangle.* URL: <http://blog.eyetools.net/>. Report available for purchase at URL: <http://www.sherpastore.com/Google-eye-tracking.html>. Last accessed May 2006.

Findlay, J. M. (1992). Programming of Stimulus-Elicited Saccadic Eye Movements. In K. Rayner (Ed.), *Eye Movements and Visual Cognition: Scene Perception and Reading* (pp. 8–30). New York: Springer-Verlag. (Springer Series in Neuropsychology)

Findlay, J. M. (1997). Saccade Terget Selection During Visual Search. *Vision Research*, *37*(5), 617–631.

Findlay, J. M., & Gilchrist, I. D. (1998). Eye Guidance and Visual Search. In G. Underwood (Ed.), *Eye Guidance in Reading and Scene Perception* (pp. 295–312). Oxford: Elsevier.

Finke, R. A. (1989). *Principles of Mental Imagery.* Cambridge, MA: MIT Press.

Fisher, D. F., Monty, R. A., & Senders, J. W. (Eds.). (1981). *Eye Movements: Cognition and Visual Perception.* Hillsdale, NJ: Lawrence Erlbaum.

Fuchs, A. F., Kaneko, C. R. S., & Scudder, C. A. (1985). Brainstem Control of Saccadic Eye Movements. *Annual Review of Neuroscience*, *8*, 307–337.

Funkhouser, T. A., & Séquin, C. H. (1993). Adaptive Display Algorithm for Interactive Frame Rates During Visualization of Complex Virtual Environments. In *Computer Graphics (SIGGRAPH '93).* New York.

Gamlin, P., & Twieg, D. (1997). *A Combined Visual Display and Eye Tracking System for High-Field FMRI Studies.* (NSF Interactive Systems Grantees Workshop (ISGW'97) Report abstract available at URL: <http://www.cse.ogi.edu/CSLU/isgw97/reports/gamlin.html>. Last accessed March 1998.)

Gazzaniga, M. S. (Ed.). (2000). *The New Neurosceinces.* Cambridge, MA: MIT Press.

Geisler, W. S., & Perry, J. S. (1998). Real-time foveated multiresolution system for low-bandwidth video communication. In *Human vision and electronic imaging.* Bellingham, WA.

Geisler, W. S., & Perry, J. S. (2002). Real-time Simulation of Arbitrary Visual Fields. In *Eye Tracking Research & Applications (ETRA) Symposium* (pp. 83–153). New Orleans.

Gibson, J. J. (1941). A Critical Review of the Concept of Set in Contemporary Experimental Psychology. *Psychological Bulletin, 38*(9), 781–817.

Glassner, A. S. (Ed.). (1989). *An Introduction to Ray Tracing.* San Diego: Academic Press.

Goldberg, J., Stimson, M., Lewnstein, M., Scott, N., & Wichansky, A. (2002). Eye Tracking in Web Search Tasks: Design Implications. In *Eye Tracking Research & Applications (ETRA) Symposium.* New Orleans.

Goldberg, J. H., & Kotval, X. P. (1999). Computer Interface Evaluation Using Eye Movements: Methods and Constructs. *International Journal of Industrial Ergonomics, 24,* 631–645.

Graeber, D. A., & Andre, A. D. (1999). Assessing Visual Attention of Pilots While Using Electronic Moving Maps for Taxiing. In R. S. Jensen, B. Cox, J. D. Callister, & R. Lavis (Eds.), *International Symposium on Aviation Psychology (ISAP)* (pp. 791–796). Columbus, OH.

Gramopadhye, A., Bhagwat, S., Kimbler, D., & Greenstein, J. (1998). The Use of Advanced Technology for Visual Inspection Training. *Applied Ergonomics, 29*(5), 361–375.

Greene, H. H., & Rayner, K. (2001). Eye Movements and Familiarity Effects in Visual Search. *Vision Research, 41*(27), 3763–3773.

Gregory, R. L. (1990). *Eye and Brain: The Psychology of Seeing.* Princeton, NJ: Princeton University Press.

Grzywacz, N. M., & Norcia, A. M. (1995). Directional Selectivity in the Cortex. In M. A. Arbib (Ed.), *The Handbook of Brain Theory and Neural Networks* (pp. 309–311). Cambridge, MA: MIT Press.

Grzywacz, N. M., Sernagor, E., & Amthor, F. R. (1995). Directional Selectivity in the Retina. In M. A. Arbib (Ed.), *The Handbook of Brain Theory and Neural Networks* (pp. 312–314). Cambridge, MA: MIT Press.

Guan, Z., Lee, S., Cuddihy, E., & Ramey, J. (2006). The Validity of the Stimulated Retrospective Think-Aloud Method as Measured by Eye Tracking. In *CHI '06: Proceedings of the SIGCHI Conference on Human Factors in Computing Systems* (pp. 1253–1262). New York: ACM Press.

Gugerty, L. (1997). Situation Awareness During Driving: Explicit and Implicit Knowledge in Dynamic Spatial Memory. *Experimental Psychology: Applied, 3*(1), 42–66.

Haber, R. N., & Hershenson, M. (1973). *The Psychology of Visual Perception.* New York: Holt, Rinehart, and Winston, Inc.

Hain, T. C. (1999). *Saccade (Calibration) Tests.* (Online Manual, URL: <http://www.tchain.com/otoneurology/practice/saccade.htm>. Last accessed October 2001.)

Harlow, J. M. (1868). Recovery from the passage of an iron bar through the head. *Publications of the Massachusetts Medical Society, 2,* 327–347.

Hayhoe, M. M., Ballard, D. H., Triesch, J., Shinoda, H., Rogriguez, P. A., & Sullivan, B. (2002). Vision in Natural and Virtual Environments (Invited Paper). In *Eye Tracking Research & Applications (ETRA) Symposium*. New Orleans.

He, P., & Kowler, E. (1989). The Role of Location Probability in the Programming of Saccades: Implications for "Center-of-Gravity" Tendencies. *Vision Research, 29*(9), 1165–1181.

Heijden, A. H. C. Van der. (1992). *Selective Attention in Vision*. London: Routledge.

Heminghous, J., & Duchowski, A. T. (2006). iComp: A Tool for Scanpath Visualization and Comparison (Poster). In *Proceedings of Applied Perception in Graphics and Visualization (APGV)*. New York: ACM Press.

Hendee, W. R., & Wells, P. N. T. (1997). *The Perception of Visual Information* (2nd ed.). New York: Springer-Verlag.

Henderson, J. M. (1992). Object Identification in Context: The Visual Processing of Natural Scenes. *Canadian Journal of Psychology, 46*(3), 319–341.

Henderson, J. M., & Hollingworth, A. (1998). Eye Movements During Scene Viewing: An Overview. In G. Underwood (Ed.), *Eye Guidance in Reading and Scene Perception* (pp. 269–294). Oxford: Elsevier.

Ho, G., Scialfa, C. T., Caird, J. K., & Graw, T. (2001). Visual Search for Traffic Signs: The Effects of Clutter, Luminance, and Aging. *Human Factors, 43*(3), 194–207.

Hoppe, H. (1997). View-Dependent Refinement of Progressive Meshes. In *Computer Graphics (SIGGRAPH '97)*. New York.

Hoppe, H. (1998). Smooth View-Dependent Level-Of-Detail Control and its Application to Terrain Rendering. In *Proceedings of IEEE Visualization* (pp. 35–42). Raleigh, NC.

Hornof, A. J., & Cavender, A. (2005). EyeDraw: Enabling Children with Severe Motor Impairments to Draw with Their Eyes. In *CHI '05: Proceedings of the SIGCHI Conference on Human Factors in Computing Systems* (pp. 161–170). New York: ACM Press.

Hornof, A. J., & Kieras, D. E. (1997). Cognitive Modeling Reveals Menu Search is Both Random and Systematic. In *Human Factors in Computing Systems: CHI '97 Conference Proceedings* (pp. 107–144). Atlanta, GA.

Hubel, D. H. (1988). *Eye, Brain, and Vision*. New York: Scientific American Library.

Hughes, H. C., Nozawa, G., & Kitterle, F. (1996). Global Precedence, Spatial Frequency Channels, and the Statistics of Natural Images. *Journal of Cognitive Neuroscience, 8*(3), 197–230.

Hutchinson, T. E. (1993). Eye-gaze computer interfaces. *IEEE Computer, 26*(7), 65,67.

Idelix Software Inc. (2004). *Pliable Display Technology White Paper v4.0*. URL: <http://www.idelix.com/pdf/pdt_whitepaper.pdf>. Last accessed December 2004.

Irwin, D. E. (1992). Visual Memory Within and Across Fixations. In K. Rayner (Ed.), *Eye movements and visual cognition: Scene perception and reading* (pp. 146–165). New York: Springer-Verlag. (Springer Series in Neuropsychology)

Itti, L., Koch, C., & Niebur, E. (1998). A Model of Saliency-Based Visual Attention for Rapid Scene Analysis. *IEEE Transactions on Pattern Analysis and Machine Intelligence (PAMI)*, *20*(11), 1254–1259.

Jacob, R. J. (1990). What You Look at is What You Get: Eye Movement-Based Interaction Techniques. In *Human Factors in Computing Systems: CHI '90 Conference Proceedings* (pp. 11–18). ACM Press.

Jacob, R. J. K., & Karn, K. S. (2003). Eye Tracking in Human-Computer Interaction and Usability Research: Ready to Deliver the Promises. In J. Hyönä, R. Radach, & H. Deubel (Eds.), *The Mind's Eye: Cognitive and Applied Aspects of Eye Movement Research* (pp. 573–605). Amsterdam, The Netherlands: Elsevier Science.

James, W. (1981). *The Principles of Psychology* (Vol. I). Cambridge, MA: Harvard University Press. (See: James, William, *The Principles of Psychology*, H. Holt and Co., New York, 1890.)

Josephson, S., & Holmes, M. E. (2006). Clutter or Content? How On-Screen Enhancements Affect How TV Viewers Scan and What They Learn. In *Proceedings of Eye Tracking Research & Applications (ETRA)* (pp. 155–162). New York: ACM Press.

Kanizsa, G. (1976). Subjective Contours. *Scientific American*, *234*(4), 48–52,138.

Kaplan, E. (1991). The Receptive Field Structure of Retinal Ganglion Cells in Cat and Monkey. In A. G. Leventhal & J. R. Cronly-Dillon (Eds.), *The Neural Basis of Visual Function* (pp. 10–40). Boca Raton, FL: CRC Press. (Vision and Visual Dysfunction Series, vol. 4)

Karn, K. S., Ellis, S., & Juliano, C. (1999). The Hunt for Usability: Tracking Eye Movements. In *Human Factors in Computing Systems: CHI '99 Conference Proceedings (Workshops)* (p. 173). ACM Press.

Kennedy, A. (1992). The Spatial Coding Hypothesis. In K. Rayner (Ed.), *Eye Movements and Visual Cognition: Scene Perception and Reading* (pp. 379–396). New York: Springer-Verlag. (Springer Series in Neuropsychology)

Kieras, D., & Meyer, D. E. (1995). *An Overview of the EPIC Architecture for Cognition and Performance with Appliaction to Human-Computer Interaction* (EPIC Tech. Rep. No. 5 No. TR-95/ONR-EPIC-5). Ann Arbor, University of Michigan, Electrical Engineering and Computer Science Department. (URL: <http://www.eecs.umich.edu/~kieras/epic.html>. Last accessed July 2006.)

Knox, P. C. (2001). *The Parameters of Eye Movement.* (Lecture Notes. URL: <http://www.liv.ac.uk/~pcknox/teaching/Eymovs/params.htm>. Last accessed October 2001.)

Kocian, D. (1987). Visual World Subsystem. In *Super Cockpit Industry Days: Super Cockpit/Virtual Crew Systems.* Air Force Museum, Wright-Patterson AFB, OH.

Koenderink, J. J., van Doorn, A. J., & van de Grind, W. A. (1985). Spatial and Temporal Parameters of Motion Detection in the Peripheral Visual Field. *J. Opt. Soc. Am.*, *2*(2), 252–259.

Kortum, P., & Geisler, W. S. (1996). Implementation of a foveated image coding system for bandwidth reduction of video images. In *Human Vision and Electronic Imaging* (pp. 350–360). Bellingham, WA.

Kosslyn, S. M. (1994). *Image and Brain.* Cambridge, MA: MIT Press.

Kroll, J. F. (1992). Making a Scene: The Debate about Context Effects for Scenes and Sentences. In K. Rayner (Ed.), *Eye Movements and Visual Cognition: Scene Perception and Reading.* Springer-Verlag. (Springer Series in Neuropsychology)

Land, M., Mennie, N., & Rusted, J. (1999). The Roles of Vision and Eye Movements in the Control of Activities of Daily Living. *Perception, 28*(11), 1307–1432.

Land, M. F., & Hayhoe, M. (2001). In What Ways Do Eye Movements Contribute to Everyday Activities. *Vision Research, 41*(25–26), 3559–3565. (Special Issue on *Eye Movements and Vision in the Natual World*, with most contributions to the volume originally presented at the 'Eye Movements and Vision in the Natural World' symposium held at the Royal Netherlands Academy of Sciences, Amsterdam, September 2000)

Laurutis, V. P., & Robinson, D. A. (1986). The Vestibulo-ocular Reflex During Human Saccadic Eye Movements. *Journal of Physiology, 373*, 209–233.

Leigh, R. J., & Zee, D. S. (1991). *The Neurology of Eye Movements* (2nd ed.). Philadelphia: F. A. Davis.

Levoy, M., & Whitaker, R. (1990). Gaze-Directed Volume Rendering. In *Computer Graphics (SIGGRAPH '90)* (pp. 217–223). New York.

Li, D., Babcock, J., & Parkhurst, D. J. (2006). openEyes: A Low-Cost Head-Mounted Eye-tracking Solution. In *Proceedings of Eye Tracking Research & Applications (ETRA)* (pp. 95–100). New York: ACM Press. (URL: <http://hcvl.hci.iastate.edu/cgi-bin/openEyes.cgi>. Last accessed May 2006.)

Lindstrom, P., Koller, D., Ribarsky, W., Hodges, L. F., Faust, N., & Turner, G. A. (1996). Real-Time, Continuous Level of Detail Rendering of Height Fields. In *Computer Graphics (SIGGRAPH '96)* (pp. 109–118). New York.

Liu, A. (1998). What the Driver's Eye Tells the Car's Brain. In G. Underwood (Ed.), *Eye Guidance in Reading and Scene Perception* (pp. 431–452). Oxford: Elsevier.

Livingstone, M., & Hubel, D. (1988). Segregation of Form, Color, Movement, and Depth: Anatomy, Physiology, and Perception. *Science, 240*, 740–749.

Loftus, G. R. (1981). Tachistoscopic Simulations of Eye Fixations on Pictures. *Journal of Experimental Psychology: Human Learning and Memory, 7*(5), 369–376.

Logothetis, N. K., & Leopold, D. A. (1995). *On the Physiology of Bistable Percepts* (AI Memo No. 1553). Artificial Intelligence Laboratory, Massachusetts Institute of Technology.

Lohse, G. L. (1997). Consumer Eye Movement Patterns on Yellow Pages Advertising. *Journal of Advertising, 26*(1), 61–73.

Longridge, T., Thomas, M., Fernie, A., Williams, T., & Wetzel, P. (1989). Design of an Eye Slaved Area of Interest System for the Simulator Complexity Testbed.

In T. Longridge (Ed.), *Area of Interest/Field-Of-View Research Using ASPT (Interservice/Industry Training Systems Conference)* (pp. 275–283). Brooks Air Force Base, TX: Air Force Human Resources Laboratory, Air Force Systems Command.

Loschky, L. C., & McConkie, G. W. (2000). User Performance With Gaze Contingent Multiresolutional Displays. In *Eye Tracking Research & Applications Symposium* (pp. 97–103). Palm Beach Gardens, FL.

Luebke, D., & Erikson, C. (1997). View-Dependent Simplification Of Arbitrary Polygonal Environments. In *Computer Graphics (SIGGRAPH '97)*. New York.

Luebke, D., Hallen, B., Newfield, D., & Watson, B. (2000). *Perceptually Driven Simplification Using Gaze-Directed Rendering* (Tech. Rep. No. CS-2000-04). University of Virginia.

Lund, J. S., Wu, Q., & Levitt, J. B. (1995). Visual Cortex Cell Types and Connections. In M. A. Arbib (Ed.), *The Handbook of Brain Theory and Neural Networks* (pp. 1016–1021). Cambridge, MA: MIT Press.

MacCracken, R., & Joy, K. (1996). Free-From Deformations With Lattices of Arbitrary Topology. In *Computer Graphics (SIGGRAPH '96)* (pp. 181–188). New York.

Majaranta, P., & Raiha, K.-J. (2002). Twenty Years of Eye Typing: Systems and Design Issues. In *Eye Tracking Research & Applications (ETRA) Symposium*. New Orleans.

Mayhew, D. J. (1999). *The Usability Engineering Lifecycle: A Practitioner's Handbook for User Interface Design*. San Diego: Academic Press.

McColgin, F. H. (1960). Movement Thresholds in Peripheral Vision. *Journal of the Optical Society of Amercia, 50*(8), 774–779.

McConkie, G. W., & Rayner, K. (1975). The Span of the Effective Stimulus During a Fixation in Reading. *Perception & Psychophysics, 17*, 578–586.

McCormick, B. H., Batte, D. A., & Duchowski, A. T. (1996). *A Virtual Environment: Exploring the Brain Forest.* (Presentation at the CENAC Conference, October 1996, Mexico City)

McNamara, A., & O'Sullivan, C. (Eds.). (2001). *Perceptually Adaptive Graphics.* Snowbird, UT: ACM SIGGRAPH/EuroGraphics. (URL: <http://isg.cs.tcd.ie/campfire/>. Last accessed May 2002.)

Megaw, E. D., & Richardson, J. (1979). Eye Movements and Industrial Inspection. *Applied Ergonomics, 10*, 145–154.

Menache, A. (2000). *Understanding Motion Capture for Computer Animation and Video Games.* San Diego: Morgan Kaufmann (Academic Press).

Microsoft Corporation. (1999). *Microsoft Windows User Experience: Official Guidelines for User Interface Developers and Designers.* Microsoft Press.

Molnar, F. (1981). About the Role of Visual Exploration in Aesthetics. In H. Day (Ed.), *Advances in Intrinsic Motivation and Aesthetics.* New York: Plenum Press.

Monty, R. A., & Senders, J. W. (Eds.). (1976). *Eye Movements and Psychological Processes.* Hillsdale, NJ: Lawrence Erlbaum.

Murphy, H., & Duchowski, A. T. (2001). Gaze-Contingent Level Of Detail. In *Eurographics (short presentations)*. Manchester, UK.

Necker, L. A. (1832). Observations on Some Remarkable Optical Phaenomena Seen in Switzerland, and on an Optical Phaenomenon Which Occurs on Viewing a Figure or a Crystal or Geometrical Solid. *Phiolosophical Magazine and Journal of Science*, *1*, 329–337. (Third Series.)

Nguyen, E., Labit, C., & Odobez, J.-M. (1994). A *ROI* Approach for Hybrid Image Sequence Coding. In *International Conference on Image Processing (ICIP)'94* (pp. 245–249). Austin, TX.

Nielsen, J. (1993). The Next Generation GUIs: Noncommand User Interfaces. *Communications of the ACM*, *36*(4), 83–99.

Nielson, J. (2006a). *F-Shaped Pattern For Reading Web Content.* URL: <http://www.useit.com/alertbox/reading_pattern.html>. Last accessed July 2006.

Nielson, J. (2006b). *Will Plain-Text Ads Continue to Rule?* URL: <http://www.useit.com/alertbox/20030428.html>. Last accessed July 2006.

Nielson Norman Group. (2006). *Eyetracking Research.* URL: <http://www.useit.com/eyetracking/>. Last accessed July 2006.

Nikolov, S. G., Newman, T. D., Bull, D. R., Canagarajah, N. C., Jones, M. G., & Gilchrist, I. D. (2004). Gaze-Contingent Display Using Texture Mapping and OpenGL: System and Applications. In *Eye Tracking Research & Applications (ETRA) Symposium* (pp. 11–18). San Antonio, TX.

Ninio, J., & Stevens, K. A. (2000). Variations on the Hermann Grid: An Extinction Illusion. *Perception*, *29*, 1209–1217.

Norman, D. A. (1988). *The Design of Everyday Things.* New York: Doubleday/Currency Ed.

Noton, D., & Stark, L. (1971a). Eye Movements and Visual Perception. *Scientific American*, *224*, 34–43.

Noton, D., & Stark, L. (1971b). Scanpaths in Saccadic Eye Movements While Viewing and Recognizing Patterns. *Vision Research*, *11*, 929–942.

Ohshima, T., Yamamoto, H., & Tamura, H. (1996). Gaze-Directed Adaptive Rendering for Interacting with Virtual Space. In *Proceedings of VRAIS'96* (pp. 103–110). Santa Clara, CA.

O'Regan, K. J. (1992). Optimal Viewing Position in Words and the Strategy-Tactics Theory of Eye Movements in Reading. In K. Rayner (Ed.), *Eye Movements and Visual Cognition: Scene Perception and Reading* (pp. 333–355). New York: Springer-Verlag. (Springer Series in Neuropsychology)

Osberger, W., & Maeder, A. J. (1998). Automatic Identification of Perceptually Important Regions in an Image. In *International Conference on Pattern Recognition*. Brisbane.

O'Sullivan, C., & Dingliana, J. (2001). Collisions and Perception. *ACM Transactions on Graphics*, *20*(3).

O'Sullivan, C., Dingliana, J., & Howlett, S. (2002). Gaze-Contingent Algorithms for Interactive Graphics. In J. Hyönä, R. Radach, & H. Deubel (Eds.), *The Mind's*

Eyes: Cognitive and Applied Aspects of Eye Movement Research. Oxford: Elsevier Science.

Ottati, W. L., Hickox, J. C., & Richter, J. (1999). Eye Scan Patterns of Experienced and Novice Pilots During Visual Flight Rules (VFR) Navigation. In *Proceedings of the Human Factors and Ergonomics Society, 43rd Annual Meeting.* Houston, TX.

Özyurt, J., DeSouza, P., West, P., Rutschmann, R., & Greenlee, M. W. (2001). Comparison of Cortical Activity and Oculomotor Performance in teh Gap and Step Paradigms. In *European Conference on Visual Perception (ECVP).* Kusadasi, Turkey.

Palmer, S. (1999). *Vision Science.* Cambridge, MA: MIT Press.

Papathomas, T. V., Chubb, C., Gorea, A., & Kowler, E. (Eds.). (1995). *Early Vision and Beyond.* Cambridge, MA: MIT Press.

Parkhurst, D., Culurciello, E., & Niebur, E. (2000). Evaluating Variable Resolution Displays with Visual Search: Task Performance and Eye Movements. In *Eye Tracking Research & Applications (ETRA)* (pp. 105–109). Palm Beach Gardens, FL.

Parkhurst, D. J., & Niebur, E. (2002). Variable Resolution Displays: A Theoretical, Practical, and Behavioral Evaluation. *Human Factors, 44,* 611–629.

Pelz, J. B., Canosa, R., & Babcock, J. (2000). Extended Tasks Elicit Complex Eye Movement Patterns. In *Eye Tracking Research & Applications (ETRA) Symposium* (pp. 37–43). Palm Beach Gardens, FL.

Peters, R. J., & Itti, L. (2006). Computational Mechanisms for Gaze Direction in Interactive Visual Environments. In *Proceedings of Eye Tracking Research & Applications (ETRA)* (pp. 27–32). New York: ACM Press.

Pirenne, M. H. (1967). *Vision and The Eye* (2nd ed.). London: Chapman & Hall.

Posner, M. I., Snyder, C. R. R., & Davidson, B. J. (1980). Attention and the Detection of Signals. *Experimental Psychology: General, 109*(2), 160–174.

Privitera, C. M., & Stark, L. W. (2000). Algorithms for Defining Visual Regions-of-Interest: Comparison with Eye Fixations. *IEEE Transactions on Pattern Analysis and Machine Intelligence (PAMI), 22*(9), 970–982.

Räihä, K.-J., Aula, A., Majaranta, P., Rantala, H., & Koivunen, K. (2005). Static Visualization of Temporal Eye-Tracking Data. In *Human-Computer Interaction–INTERACT* (pp. 946–949). New York. (Lecture Notes in Computer Science 3585)

Rayner, K. (1975). The Perceptual Span and Peripheral Cues in Reading. *Cognitive Psychology, 7,* 65–81.

Rayner, K. (Ed.). (1992). *Eye Movements and Visual Cognition: Scene Perception and Reading.* New York: Springer-Verlag. (Springer Series in Neuropsychology)

Rayner, K. (1998). Eye Movements in Reading and Information Processing: 20 Years of Research. *Psychological Bulletin, 124*(3), 372–422.

Rayner, K., & Bertera, J. H. (1979). Reading Without a Fovea. *Science, 206,* 468–469.

Rayner, K., & Pollatsek, A. (1992). Eye Movements and Scene Perception. *Canadian Journal of Psychology*, *46*(3), 342–376.

Rayner, K., Rotello, C. M., Stewart, A. J., Keir, J., & Duffy, S. A. (2001). Integrating Text and Pictorial Information: Eye Movements When Looking at Print Advertisements. *Journal of Experimental Psychology: Applied*, *7*(3), 219–226.

Recarte, M. A., & Nunes, L. M. (2000). Effects of Verbal and Spatial-Imagery Tasks on Eye Fixations While Driving. *Journal of Experimental Psychology: Applied*, *6*(1), 31–43.

Reddy, M. (1998). Specification and Evaluation of Level of Detail Selection Criteria. *Virtual Reality: Research, Development and Application*, *3*(2), 132–143.

Reddy, M. (2001). Perceptually Optimized 3D Graphics. *Computer Graphics and Applications*, *21*(5), 68–75.

Reingold, E. M., Charness, N., Pomplun, M., & Stampe, D. M. (2002). Visual Span in Expert Chess Players: Evidence from Eye Movements. *Psychological Science*, *12*(1), 48–55.

Reingold, E. M., Loschky, L. C., McConkie, G. W., & Stampe, D. M. (2002). Gaze-Contingent Multi-Resolutional Displays: An Integrative Review. *Human Factors*, *45*(2), 307–328.

Rele, R., & Duchowski, A. (2005). Using Eye Tracking to Evaluate Alternative Search Results Interfaces. In *Proceedings of the Human Factors and Ergonomics Society, 49th Annual Meeting*. Orlando, FL.

Renshaw, T., Webb, N., & Finlay, J. (2006). Geatting a Measure of Satisfaction from Eyetracking in Practice. In *Human Factors in Computing Systems: CHI '06 Conference Proceedings (Workshops)*. ACM Press.

Robinson, D. A. (1968). The Oculomotor Control System: A Review. *Proceedings of the IEEE*, *56*(6), 1032–1049.

Rosbergen, E., Wedel, M., & Pieters, R. (1990). *Analyzing Visual Attention to Repeated Print Advertising Using Scanpath Theory* (Tech. Rep.). University Library Groningen, SOM Research School. (# 97B32)

Sadasivan, S., Greenstein, J. S., Gramopadhye, A. K., & Duchowski, A. T. (2005). Use of Eye Movements as Feedforward Training for a Synthetic Aircraft Inspection Task. In *CHI '05: Proceedings of SIGCHI Conference on Human Factors in Computing Systems* (pp. 141–149). New York: ACM Press.

Salvucci, D. D., & Anderson, J. R. (2001). Automated Eye-Movement Protocol Analysis. *Human-Computer Interaction*, *16*, 39–86.

Salvucci, D. D., & Goldberg, J. H. (2000). Identifying Fixations and Saccades in Eye-Tracking Protocols. In *Eye Tracking Research & Applications (ETRA) Symposium* (pp. 71–78). Palm Beach Gardens, FL: ACM.

Santella, A., Agrawala, M., DeCarlo, D., Salesin, D., & Cohen, M. (2006). Gaze-Based Interaction for Semi-Automatic Photo Cropping. In *CHI '06: Proceedings of the SIGCHI Conference on Human Factors in Computing Systems* (pp. 781–790). New York: ACM Press.

Schmalstieg, D., & Schaufler, G. (1997). Smooth Levels of Detail. In *Proceedings of VRAIS'97* (pp. 12–19). Albuquerque, NM.

Schoonard, J. W., Gould, J. D., & Miller, L. A. (1973). Studies of Visual Inspection. *Ergonomics*, *16*(4), 365–379.

Schroeder, W. E. (1993a). Head-mounted computer interface based on eye tracking. In *Visual Communications and Image Processing '93 (VCIP)* (pp. 1114–1124). Bellingham, WA.

Schroeder, W. E. (1993b). Replacing mouse and trackball with tracked line of gaze. In *Visual Communications and Image Processing '93 (VCIP)* (pp. 1103–1113). Bellingham, WA.

Shebilske, W. L., & Fisher, D. F. (1983). Understanding Extended Discourse Through the Eyes: How and Why. In R. Groner, C. Menz, D. F. Fisher, & R. A. Monty (Eds.), *Eye Movements and Psychological Functions: International Views* (pp. 303–314). Hillsdale, NJ: Lawrence Erlbaum.

Shell, J. S., Vertegaal, R., Cheng, D., Skaburskis, A. W., Sohn, C., & Stewart, A. J. (2004). ECSGlasses and EyePliances: Using Attention to Open Sociable Windows of Interaction. In *Proceedings of Eye Tracking Research & Applications (ETRA)* (pp. 93–100). New York: ACM Press.

Shepherd, S., Schmitt, D., & Platt, M. (2004). Social Orienting by Primates Moving Freely in Naturalistic Environments (Abstract Viewer/Itinerary Planner). In *Program No. 332.3*. Society for Neuroscience. (URL: <http://www.duke.edu/~svs/research/>. Last accessed May 2006.)

Shumberger, M., Duchowski, A. T., & Charlow, K. (2005). Human Factors Engineering in Command, Control & Intelligence (C2I) User Interface Development Process: Use of Eye Tracking Methodology. In *Human Systems Integration Symposium*. Arlington, VA.

Sibert, L. E., & Jacob, R. J. (2000). Evaluation of Eye Gaze Interaction. In *Human Factors in Computing Systems: CHI 2000 Conference Proceedings*. ACM Press.

Smeets, J. B. J., Hayhoe, H. M., & Ballard, D. H. (1996). Goal-Directed Arm Movements Change Eye-Head Coordination. *Experimental Brain Research*, *109*, 434–440.

Smith, J. D., Vertegaal, R., & Sohn, C. (2005). ViewPointer: Lightweight Calibration-Free Eye Tracking for Ubiquitous Handsfree Deixis. In *UIST '05: Proceedings of the 18th Annual ACM Symposium on User Interface Software and Technology* (pp. 53–61). New York: ACM Press.

Snodderly, D. M., Kagan, I., & Gur, M. (2001). Selective Activation of Visual Cortex Neurons by Fixational Eye Movements: Implications for Neural Coding. *Visual Neuroscience*, *18*, 259–277.

Solso, R. L. (1999). *Cognition and the Visual Arts* (3rd ed.). Cambridge, MA: MIT Press.

Sparks, D. L., & Mays, L. E. (1990). Signal Transformations Required for the Generation of Saccadic Eye Movements. *Annual Review of Neuroscience*, *13*, 309–336.

Starker, I., & Bolt, R. A. (1990). A Gaze-Responsive Self-Disclosing Display. In *Human Factors in Computing Systems: CHI '90 Conference Proceedings* (pp. 3–9). ACM Press.

Stelmach, L. B., & Tam, W. J. (1994). Processing Image Sequences Based on Eye Movements. In *Conference on Human Vision, Visual Processing, and Digital Display V* (pp. 90–98). San Jose, CA.

Stone, D., Jarrett, C., Woodroffe, M., & Minocha, S. (2005). *User Interface Design and Evaluation*. San Francisco: Morgan Kaufmann Publishers.

Tanriverdi, V., & Jacob, R. J. K. (2000). Interacting with Eye Movements in Virtual Environments. In *Human Factors in Computing Systems: CHI 2000 Conference Proceedings* (pp. 265–272). ACM Press.

Tarasewich, P., Pomplun, M., Fillion, S., & Broberg, D. (2005). The Enhanced Restricted Focus Viewer. *International Journal of Human-Computer Interaction*, *19*(1), 35–54.

Tobii Technology AB. (2003). *Tobii ET-17 Eye-tracker Product Description*. (Version 1.1)

Todd, S., & Kramer, A. F. (1993). Attentional Guidance in Visual Attention. In *Proceedings of the Human Factors and Ergonomics Society, 37th Annual Meeting* (pp. 1378–1382). Santa Monica, CA.

Tole, J. R., & Young, L. R. (1981). Digital Filters for Saccade and Fixation Detection. In D. F. Fisher, R. A. Monty, & J. W. Senders (Eds.), *Eye Movements: Cognition and Visual Perception* (pp. 185–199). Hillsdale, NJ: Lawrence Erlbaum.

Treisman, A. (1986). Features and Objects in Visual Processing. *Scientific American*, *255*(5), 114B–125,140.

Treisman, A., & Gelade, G. (1980). A Feature Integration Theory of Attention. *Cognitive Psychology*, *12*, 97–136.

Tsumura, N., Endo, C., Haneishi, H., & Miyake, Y. (1996). Image compression and decompression based on gazing area. In *Human Vision and Electronic Imaging*. Bellingham, WA.

Underwood, G. (Ed.). (1998). *Eye Guidance in Reading and Scene Perception*. Oxford: Elsevier Science.

Uwano, H., Nakamura, M., Monden, A., & Matsumoto, K.-i. (2006). Analyzing Individual Performance of Source Code Review Using Reviewers' Eye Movement. In *Eye Tracking Research & Applications (ETRA) Symposium* (pp. 133–140). San Diego.

Van Orden, K. F., & DiVita, J. (1993). Highlighting with Flicker. In *Proceedings of the Human Factors and Ergonomics Society, 37th Annual Meeting* (pp. 1300–1304). Santa Monica, CA.

Velichkovsky, B., Pomplun, M., & Rieser, J. (1996). Attention and Communication: Eye-Movement-Based Research Paradigms. In W. H. Zangemeister, H. S. Stiehl, & C. Freksa (Eds.), *Visual Attention and Cognition* (pp. 125–154). Amsterdam, Netherlands: Elsevier Science.

Vembar, D., Iyengar, N., Duchowski, A. T., Clark, K., Hewitt, J., & Pauls, K. (2004). Effect of Visual Cues on Human Performance in Navigating Through a Virtual Maze. In *EuroGraphics Symposium on Virtual Environments (EGVE)*. Grenoble, France.

Vertegaal, R. (1999). The GAZE Groupware System: Mediating Joint Attention in Mutiparty Communication and Collaboration. In *Human Factors in Computing Systems: CHI '99 Conference Proceedings* (pp. 294–301). ACM Press.

Vertegaal, R. (2003). Attentive User Interfaces (Editorial, Special Issue on Attentive User Interfaces). *Communications of the ACM, 46*(3).

Vince, J. A. (1995). *Virtual Reality Systems.* Reading, MA: Addison-Wesley Longman.

Von Helmholtz, H. (1925). *Handbuch der Physiologischen Optik (Treatise on Physiological Optics)* (Vol. III, Translated from the Third German ed.). Rochester, NY: The Optical Society of America.

Wallace, G. K. (1991). The JPEG Still Picture Compression Standard. *Communications of the ACM, 34*(4), 30–45.

Wang, M.-J. J., Lin, S.-C., & Drury, C. G. (1997). Training for strategy in visual search. *Industrial Ergonomics, 20,* 101–108.

Waterman, M. S. (1989). Sequence Alignments. In M. S. Waterman (Ed.), *Mathematical Methods for DNA Sequences* (pp. 53–92). Boca Raton, FL: CRC Press.

Watson, B., Walker, N., & Hodges, L. F. (1997). Managing Level of Detail through Head-Tracked Peripheral Degradation: A Model and Resulting Design Principles. In *Virtual Reality Software & Technology: Proceedings of the VRST'97* (pp. 59–63). ACM.

Watson, B., Walker, N., & Hodges, L. F. (2004). Supra-Threshold Control of Peripheral LOD. *Transaction on Graphics (SIGGRAPH'04 Proceedings), 23*(3), 750–759.

Watson, B., Walker, N., Hodges, L. F., & Worden, A. (1997). Managing Level of Detail through Peripheral Degradation: Effects on Search Performance with a Head-Mounted Display. *ACM Transactions on Computer-Human Interaction, 4*(4), 323–346.

Webb, N., & Renshaw, T. (2006). Adding Value with Eyetracking. In *Proceedings of UPA '05 (Workshops).* Broomfield, CO.

Wedel, M., & Pieters, R. (2000). Eye Fixations on Advertisements and Memory for Brands: A Model and Findings. *Marketing Science, 19*(4), 297–312.

West, J. M., Haake, A. R., Rozanski, E. P., & Karn, K. S. (2006). eyePatterns: Software for Identifying Patterns and Similarities Across Fixation Sequences. In *Proceedings of Eye Tracking Research & Applications (ETRA)* (pp. 149–154). New York: ACM Press.

Wolfe, J. M. (1993). Guided Search 2.0: The Upgrade. In *Proceedings of the Human Factors and Ergonomics Society, 37th Annual Meeting* (pp. 1295–1299). Santa Monica, CA.

Wolfe, J. M. (1994). Visual Search in Continuous, Naturalistic Stimuli. *Vision Research, 34*(9), 1187–1195.

Wolfe, J. M., & Gancarz, G. (1996). GUIDED SEARCH 3.0: A Model of Visual Search Catches Up With Jay Enoch 40 Years Later. In V. Lakshminarayanan (Ed.), *Basic and Clinical Applications of Vision Science* (pp. 189–192). Dordrecht, Netherlands: Kluwer Academic.

Wooding, D. (2002). Fixation Maps: Quantifying Eye-Movement Traces. In *Eye Tracking Research & Applications (ETRA) Symposium.* New Orleans.

Yarbus, A. L. (1967). *Eye Movements and Vision.* New York: Plenum Press.

Young, L. R., & Sheena, D. (1975). Survey of Eye Movement Recording Methods. *Behavior Research Methods & Instrumentation, 7*(5), 397–439.

Zee, D. S., Optican, L. M., Cook, J. D., Robinson, D. A., & Engel, W. K. (1976). Slow Saccades in Spinocerebellar Degeneration. *Archives of Neurology, 33,* 243–251.

Zeki, S. (1993). *A Vision of the Brain.* Osney Mead, Oxford: Blackwell Scientific.

Zhai, S., Morimoto, C., & Ihde, S. (1999). Manual and Gaze Input Cascaded (MAGIC) Pointing. In *Human Factors in Computing Systems: CHI '99 Conference Proceedings* (pp. 246–353). ACM Press.

Zorin, D., & Schröder, P. (2000). *Course 23: Subdivision for Modeling and Animation.* New York. (SIGGRAPH 2000 Course Notes, URL: <http://www.mrl.nyu.edu/dzorin/sig00course/>. Last accessed December 2000.)

Zorin, D., Schröder, P., & Sweldens, W. (1997). Interactive Multiresolution Mesh Editing. In *Computer Graphics (SIGGRAPH '97).* New York.

Index

Printed in the United States of America